# NOT READING HERAKLEITOS

*Other books by Alan Loney*

*& the Ampersand*, Black Light Press, Wellington 1990
*The Falling: a memoir*, Auckland University Press, New Zealand 2001
*Bruno Leti, Survey, Artists Books 1982–2003*, Geelong Gallery, Australia 2003
*The Printing of a Masterpiece*, Black Pepper Publishing, Melbourne 2008
*Each new book*, Code(x)+1 No.3, The Codex Foundation, California 2008
*The books to come*, Cuneiform Press, Houston, Texas 2010
*Bruno Leti : Paintings*, Macmillan Mini-Art Series #17, Melbourne 2011
*Anne of the Iron Door* (novella), Black Pepper Publishing, Melbourne 2011
*Beginnings*, Otis Books | Seismicity Editions, Los Angeles 2016
*In Search of the Book as a Work of Art*, Opifex, Sydney 2019


*Recent poetry*

*Day's eye*, Rubicon Press, Edmonton 2008
*Fishwork* (with Max Gimblett), The Holloway Press, Auckland 2009
*Katalogos*, Red Dragonfly Press, Minnesota 2010
*The Sirens*, Ninja Press, California 2011
*LOOM* (with Richard Wagener), Nawakum Press, California 2013
*Crankhandle*, Cordite Books, Melbourne 2015
*Melbourne Journal*, UWA Publishing, Perth 2016
*Heidegger's bicycle*, Paekakariki Press, Walthamstowe 2017
*Next to nothing*, Red Dragonfly Press, Minnesota 2018
*49 Days*, Greenboathouse Press, BC, Canada 2018

Alan Loney

# NOT READING HERAKLEITOS

foreword by Edward Jenner

re.press | Melbourne | 2021

## re.press

http://www.re-press.org

Foreword copyright © 2021 Edward Jenner
Text copyright © 2021 Alan Loney

The monoline Greek type on pages 34–35 is Diogenes Pro, designed by Christopher Stinehour and used here with his kind permission.

A catalogue record for this
book is available from the
National Library of Australia

ISBN  978-0-6487282-1-4

Series: Transmission

Cover art: WORD, glass, 4x4x2 inches, Pamela Stadus
Cover photograph: Albert Comper
Book design: Alan Loney
Typesetting: Nick Summers

# Contents

# Acknowledgments

I owe more than I can say to Randall McLeod at University of Toronto for the overall sense and aim of these notes, for the intellectual rigor of his writings, for the notion of 'not reading' Herakleitos, and his generosity in photographing many pages from the first printed collection of preSocratic philosophers by Henri Estienne, Geneva, in 1573 – to Peter Koch in Berkeley, California, I am indebted to his own continuing engagement with the preSocratics, his published editions of Diogenes, Herakleitos and Parmenides, along with his down-to-earth approach to the whole arc of ancient wisdom over the last 2500 years – I am forever grateful to Paul F Gehl (curator emeritus, Newberry Library, Chicago) for his acute reading of the raw first draft, both in much of its detail and of its structure – to Francis McWhannell (independent researcher in New Zealand) I owe thanks for his forensic copy-edit & several pertinent observations on the text – a reading of the second draft by Bryan Cooke (University of Melbourne) corrected a number of errors and enabled me to rethink parts of the essay afresh – here also I record a most pleasant and helpful email conversation with Stephen V Tracy, Professor Emeritus, Ohio State University, on how certain errors in ancient Greek inscriptions were made, and how difficult are the questions around ancient literacy – and to Justin Clemens (University of Melbourne) and Paul Ashton (Victoria University) at *re.press books* for their enthusiasm for the project and their subsequent encouragement and support during all stages of the writing – and to Ted Jenner, friend, poet, classicist, whose word has been of first importance to me, I am indebted beyond what can be repaid

# Preface

this work is personal – it makes no claim to scholarly status or originality, but it records an effort by a nonexpert to come to grips with a field in which he has a long-standing interest and in which the available works are almost all written for other experts – my interest in Herakleitos began in the 1960s with Martin Heidegger's *Being and Time,* and Kathleen Freeman's *Ancilla to the Pre-Socratic Philosophers* – I still have these volumes, altho both are a bit battered with the years – since then I have come to the sense that Herakleitos was one of the finest innovative authors who lived – such authors appear only occasionally thruout history, and most of them are not typical of their own time – it will seem that a large claim is being made on small evidence, and that is true – and honestly, I cannot make anything like a convincing case that anyone else should accept – all that's possible is to say that it seems to me to be so on the basis of a lifetime of reading a limited amount of poetry and of philosophy, east and west, and writing a few lines on paper here and there in answer to what poet Louis Zukofsky called 'deep need' –

these meditations hope to uncover a non-expert approach to what an ancient Greek philosopher said, how I can learn to read it, what are the marks on what writing surface I should accept as his, and to figure out how a general reader can find their way thru a labyrinth of academic and scholarly apparatuses and protocols and sit down as it were and have a decent chat with this venerable curmudgeon who developed ways of writing/talking unknown before him, and barely registered after him except by Martin Heidegger in some respects and by Eric A Havelock in other respects – don't get me wrong : I make no claim to have achieved any such intimacy as 'a decent chat' – all I can say is that I think I found an address where he might have lived – who knows –

there is no direct access to Herakleitos, & anyone who wishes to approach him must inevitably depend on the work of others, & the scholarship and translations of T M Robinson, M Marcovich, Guy Davenport, G S Kirk, Charles H Kahn, H Diels, I Bywater/G T W Patrick, Kathleen Freeman, Andrei Lebedev, have been the basis of my thinking and searching into 'many things' – part of my impulse to undertake this process was an interest in printing history as well as an interest in the preSocratics, and a number of essays by bibliographers & textual critics (D F McKenzie, Jerome J McGann, Randall McLeod, G Thomas Tanselle, among others) – much of which, as Keats once put it, 'dovetailed in my mind', setting me forth on this, even for me, unusual quest – I have Ted Jenner to thank for the privilege of printing his texts & translations of Ibykos at The Holloway Press back in 1997 – his *The Love Songs of Ibykos* remains an exemplum for me as is Anne Carson's *Sappho* and Guy Davenport's *Archilochos* –

other guides for me have been the writings of philosophers Martin Heidegger, A N Whitehead, Hans-Georg Gadamer and Ludwig Wittgenstein, poet Charles Olson, essayist Guy Davenport, and classicist Jane Ellen Harrison – not as commentators on Herakleitos only, but as those who have taught me how to think in a world where Opinion (said Parmenides) reigns everywhere one looks and listens – along with Wittgenstein, I read (or 'do') philosophy and read (or 'do') poetry so I can learn how to live – so yes, I absolutely do blame them for my shortcomings…

*Melbourne 2021*

'Not reading Herakleitos'? The title is doubly intriguing if you are aware
that Alan Loney describes himself as an amateur in the field of Classical
literature and maintains that his 'main interest in books is, everything aside
from reading'. Even reading a book often strikes him as a 'presumption of
power, a kind of violence, to read the words of another in their absence'.
Obviously, I'm presuming that Alan does read, and Greek sometimes too.
But what if that 'another' he mentions has been dead for at least 2500 years,
and most probably published nothing that we would recognise as a book?
Alan's scruples regarding reading would be compounded, would they not?

Disregard the shape or form the work of this ancient philosopher might
have assumed, it remains a mental construct, almost a Jungian archetype,
a 'livre spirituel' somewhat like Mallarmé's proposition, the volume that is
'l'hymne, harmonie et joie, des relations entre tout' (the hymn, harmony &
joy, of the connections between all things). The French poet's description is
a little reminiscent of the Herakleitean logos, which has traditionally been
defined as a principle of measured change and proportional interaction
between the elements of the universe, or less abstractly, as elements in a
self-regulating process (harmoniē) which maintains the delicate balance
of nature.

The contents of both 'books' would surely appear to an averagely intelligent reader as something like Mallarmé's description of his own 'book':
'a pure unity gathered in some dazzling circumstance' – what a pity no one
will ever read them! – the one being a poet's 'instrument spirituel', the other
the semi-legendary site of a ruin, remnants of which have been quarried to
bolster this philosophical argument or refute that untenable premise. But
Alan Loney has little interest in a 'livre spirituel', or a Platonic idea or any

archetype regarding the book – or at least so I surmise from a reading of this volume and from his own disquisition on the book, *The books to come* (2012), from which the quotations in the first paragraph have been taken. He believes that what Herakleitos wrote was nothing we would ever think of as a book, and approaches the ancient Pre-Socratic with his characteristic quasi-inquisitional style of meditation. This writer is a born worrier – he worries a point, chafes and frets at it until he is happy that it cannot be vexed any further – for the moment.

In his 'preamble' Alan quickly gets down to the urgent business of interrogation. Herakleitos' statements are in fragments, but what is a fragment? Why is the word 'aphorism' (a word I have used myself in relation to Herakleitos) inappropriate? Are these statements quotations? ('All we have are quotations by those who had their own motives for doing so', p2). Is the word 'sayings' a more useful term? And how are we supposed to spell his name, Herakleitos or Heraclitus?

Given the amount of phonetic effects in what remains of Herakleitos – puns, assonance, antithesis, symmetry – it is difficult not to conclude with Eric Havelock that 'statements of this type were framed not to be read but to be heard and memorized' (p45). Needless to say, our philosopher, composing in the late 5th century BC, belongs to a predominantly oral culture. Furthermore, as someone (Alan?) has calculated, in about 10% of Herakleitos, the logos is heard and listened to, in other words auditory perception is paramount (logos is actually cognate with legein, 'to say'). Thus the term 'sayings' is more appropriate than 'writings' or 'fragments'? Agreed, but though it is somewhat easy to imagine these 'sayings' being read out to a small audience, we have not the slightest idea how they got recorded and ended up ringing in the ears of a Socrates.

Scholar and philosopher Eva Brann thinks that Herakleitos' 'sayings' constitute examples of the first genre of philosophy, 'the thought-compacted aphorism'. But she is treating the 'sayings' as more or less complete statements, some of which may have been part of a much longer statement. The publication of *The Derveni Papyrus on Heraclitus* (2011) indicates that at least some of our quotations have been excerpted from a much more complicated context. Fr.3 ('the sun's breadth is the length of a human foot') now seems to have accompanied fr.94 ('the sun will not overstep his measures; otherwise the ministers of Justice, the Erinyes, will find him out'). Isn't the first something about the deceptiveness of appearances (unless it is a characteristic jibe at scientific dogmatism)? Perhaps, but either way it seems to contaminate the second which is otherwise a figurative expression of the logos, the principle of measured change, is it not? The Derveni quotation is disconcerting, mingling in the one statement two apparently discrepant 'sayings'. Alan's conclusion is all too just:

it may be an exceptional case, we don't know, but it is definitely some sort of index of the extraordinary fragility, not only of the physical evidence, but also of the readings that can be made of it. (p57)

Whatever 'book' or 'work' Herakleitos wrote (but haven't we laid that ghost to rest?), we may doubt that it was a unified account of the cosmos with neat subject headings of the sort we find in the modern histories of Pre-Socratic philosophy. It may well have been a jumble of haphazard notes rather like, as Alan suggests, Aulus Gellius' *Attic Nights*, Leonardo da Vinci's notebooks, or Alan's own notebook poems which betray more of a dread of established metre than Herakleitos who occasionally slipped (?) into hexameters (as in fr.3), which partly accounts for Eva Brann's judgement of his language and style – 'a prose that contends with poetry'.

With no primary sources, & statements that are quotations in the works of others, it is glaringly obvious that reading Herakleitos is not going to be as straight-forward as reading Plato or Aristotle. As Alan says with dry humour, 'one can only join in the chatter that has been in progress for centuries' (p11). Some quotations inevitably entail questions of authenticity: are those the words Herakleitos actually used or is this sentence a paraphrase? We might expect the quotations of the Christian apologist Hippolytos to be suspect, and yet he is regarded by reputable scholars (Kirk, Raven and Schofield) as a 'reliable source of verbatim quotation'.

Some 'sayings' with their contested words or phrases seem to be as stable as radioactive isotopes. Eight Greek words, say, might engender ten times that word count in an erudite commentary. With some justification, Alan complains that often in the learned editions of the philosopher's 'sayings', scholarly opinion not only overwhelms but sometimes even replaces discussion about the actual Greek text. Furthermore,

many of the editions, the translations, the commentaries…seek to elaborate on what the texts ought, in their view, to mean, rather than let them speak no further than what the texts can be seen to say (p21)

And yet Alan Loney, a former publisher, typesetter, and editor, is forced to admit that an interest in the actual letters on the page 'can only be at the cost of what can be understood'. So he joins his fellow Herakleiteans in an attempt to establish a text (Diels fr.59) which resists such disrespectful treatment. Perhaps it is safest to assume Dr Johnson's position. Johnson said of the risks involved in emendation, 'I have adopted the Roman sentiment, that it is more honourable to save a citizen than to kill an enemy'. But what if we are not too sure of the identity of the citizens we are supposed to save?

There is some discussion in depth in this book of just how difficult it is to get back to a quotation of Herakleitos earlier than the late 19th century editions of Diels and Bywater. Grand claims were made for Estienne and Scaliger's *Poesis Philosophica* (1573), claims which Alan examines with characteristic rigour in a chapter humorously entitled 'ad-break'. This foundational anthology of Pre-Socratic philosophers appears to have exerted next to no influence on modern editions or translations. And the consequence of such neglect is our 'unwarranted certainties of text, context, and interpretation'. The reason why such primary sources are so difficult to access, as Alan acknowledges later (when the 'penny hath dropped'), is that only a very small number of scholars is privileged enough to consult them.

What are the barriers to engaging with Herakleitos for the neophyte in Greek with no Latin? Only printed resources in English are at Alan's disposal. The lack of Latin precludes him from studying many a critical apparatus. Even abbreviations in an edition of the Pre-Socratics in the Loeb Classical Library (their avowed aim being to make the Classics available to the general public) seem to indicate that the scholars are writing for each other rather than the hoi polloi, among whose number Alan modestly counts himself ('hoi polloi, c'est moi'). Herakleitos himself had a derogatory comment concerning the 'idiots' who are 'in a flutter at every word', but being 'in a flap' over words is part of what Alan calls 'the scholarly deal, the intellectual contract' of all who write in a specialised area, no matter how extensive their knowledge of Latin and Greek.

Without a deep knowledge of the ancient tongue, Alan confesses that he cannot detect the various allusions to past and contemporary poets and philosophers in Herakleitos' 'sayings', nor can he read a complete article in Liddell & Scott's *Greek-English Lexicon*. Instead, what he brings to this ironically entitled work are the strengths he shows in all his essays, namely an enquiring mind and the clarity and frankness of an articulate intellectual who has found a congenial if frustratingly complex subject to write about. What is particularly valuable about this book, however, is its bibliographical approach to ancient texts and their transmission, an approach which is all too often overlooked by contemporary scholars. A background in printing and a fascination with the book as a cultural artefact have, I believe, served our author very well indeed. This is a book which will no doubt prove to be invaluable for all graduates embarking on a study of any branch of literature, the texts of which stem from ancient handwritten documents. Are they to rely solely (as did I, even as a postgraduate) on the sometimes dubious authority of the modern editor and their selective critical apparatus?

we are very conscious of the difference between
prose and poetry…but this distinction is not
made in the early Greek world
    *Eleanor Irwin*

what we say, he cannot understand
   what he says, we do not say
      *Han Shan, via Arthur Waley*

and in a flutter at every word
      *Herakleitos, via Kathleen Freeman*


      *a slight loosening*
    *a slight tightening*
     *of the quoins*

in memoriam Ted Jenner 1946–2021

> for the poet there is no other way
> than by his language
> *Charles Olson*

◆   Herakleitos does not say what ὁ λόγος (ho logos/the word, the account), is –
he says there is such an account, it is the same for all, it is one, it is
many, maintains its balance by 'conflict' and 'justice', and its other name is,
or may be, ὁ νόμος, (ho nomos/the law or custom), ὁ κόσμος, or any of a
number of names –
if I put this to any physicist or astrophysicist, and asked what they'd
take ὁ λόγος to mean, they might say something like 'quantum physics' –
Herakleitos was around at the birth, not only of philosophical
thought, but also of scientific thought – and that puts him as one who was
present at the rise of an account that can be valid for all, independently of
their personal opinions and beliefs –
the question is not what Herakleitos 'had in mind' or 'intended' (we
can't know that), but what kind of account fits his description

◆   wonderful link by Charles H Kahn : 'How can I be the object of my own
search? This will make sense only if my self…is difficult to find…self-knowl-
edge is difficult because a man is divided from himself' – it links up 'I went in
search of myself' & 'men…are at odds with that which they most constantly
associate',[1] (tho I prefer Charles Olson's '…estranged from that which was
most familiar',[2] –

always, in the sayings, a sense of distance between self & world, self & truth,
self and logos – 'most men do not think things in the way they encounter
them…'[3]

question is, How to think Herakleitos, when no writings survive, & (listening to Gadamer, all we have are quotations by those who had their own motives for doing so – even the dictionary of ancient Greek we have/use was compiled by christian clerics, Henry George Liddell and Robert Scott, in the language of Victorian England

♦  how to write on ὁ λόγος as if no post-Platonic work on it exists

> listen, not to me but to the λόγος –
> > are we embedded
> > > in a cosmos that, even so
> > can be thought

> > did Herakleitos live/work at a point
> > > where thought was freeing itself
> > > > from custom and opinion –

♦  how to write on ὁ λόγος as if no post-christian work on it exists

that is particularly difficult to do when a number of recent commentators have insisted on translating ὁ θεός (ho theos/the god) as 'God' instead of 'god' or 'the god' – the initial capital, which was not visible in ancient Greek writing, marks the term as christian, and therefore should not be used for anything in a text by Herakleitos –

♦  could there be a way in which this enterprise is about words, & nothing else – about what might have been the words of Herakleitos, certainly – but what could/should be the words that any of us could reasonably utter or write or remember in answer to them – are there words that have a finite life, not in the sense that they slide out of community use ('obs.', as the dictionaries tell us about some words or some uses of them), but in the sense that they can no longer open us to the life we wish for, or the life we thought we had, or even did have – might some of the words we use be dead in our mouths while we still live – what possible use could a word like 'God' or 'religion' or 'spiritual', whatever meaning anyone wishes to give them, be in my own atheistic specific time/place/comprehension as I wish to understand the words of another, not as they intended, hoped for, or even wanted to insist upon, but simply as I can speak, in my response to what Herakleitos noted as my 'encounter' with the things of the world – we are, are we not, apt to imagine that, if we use a word, then it automatically *refers*, automatically attaches to something tangible in the world, something we could get together and point at, and agree about – but what if that is not always the case – what if various words we want to use are exhausted, the culture has used them up – that have in fact used us, and worse, at the behest of those who wish to extend some measure of power over what we are and what we do and what we say – many of us, of course, are vividly aware of that – but what steps have we taken to protect ourselves from such decay, such verbal dilapidation, such shabbiness of meaning – after all, we already chatter enough with each other, don't we –

it is not that one needs to construct an argument against their use, or against their supposed 'reality', but rather to refuse to use them, and see what, if anything, is lost by such refusal – there are many words still operative in the culture that just do not call to or answer to any need or interest that I have or feel or think or perceive, see, or hear – the prospect thus presented is that nothing at all will be lost, and it's certainly logically possible that something could be gained

♦   it is customary to use the term 'fragment' for each of the sayings of the preSocratic philosophers including Herakleitos – but what is a 'fragment', and how do we distinguish it from other categories of verbal word clusters – what can we mean when we say 'fragment' in relation to any writing at all – for some time the word 'fragment' has acquired a sort of multi-layered patina to which sundry dimensions of literal and figurative meanings have been attached, making a sign of deceptively indeterminate scope – I have myself been lulled into calling many of my own writings 'fragments' when it is clear that each little cluster of words I compose is actually complete and fairly self-contained – I now, after many years, even decades, believe it's not possible for me to deliberately write in 'fragments' –

a large amount of ancient Greek writing comes to us on pieces of papyrus, most of which are not whole, but partial – damaged shreds and scraps of papyrus with partial words and partial letters preserved in large numbers – the papyrus document is a fragment, and its text an incomplete portion of a whole text – the SOED has 'fragment' as 'a piece broken off' as a literal meaning, with 'a detached, isolated, or incomplete part' of a text as a figurative meaning – in this sense, the use of 'fragment' referring to both papryus fragments and their incomplete texts is a continual sliding back and forth between literal and figurative uses of the word –

when it comes to Herakleitos, very little of his work is preserved on extant papyrus, and what is is confined to one small passage quoted in the Derveni Papyrus, found in the rubbish of a funeral pyre in 1962, in which the papyrus itself is fragmented, and in which the Herakleitean text is incomplete –

all other examples of the 'writings' of Herakleitos come to us as part of the writings of other people, who are either directly quoting or paraphrasing him – so, can we call these quotations and references 'fragments' – for if we can, then the lexical use we are making of the term 'fragment' is a figurative one, not a literal one – if instead we used the term 'quotation', then we are using a literal term rather than a figurative one – 'quotation' is literal, 'fragment' is figurative when it comes to the incomplete texts of Herakleitos and the other preSocratics –

some, however, have used 'aphorism' for the short sayings of Herakleitos – SOED has 'aphorism' as 'a 'definition or concise statement of a principle in any science', and 'Any principle or precept expressed shortly and pithily' – either way, one of the characteristics of an aphorism as defined here is completeness – a few words = the whole message – but this is not the accepted understanding of the preSocratic texts we have – to use the term 'aphorism' in this sense of any Herakleitean saying, we need to know that the quotation we are citing is not just a genuine saying, but is also the *complete* saying, with nothing lost or yet to be found – referring back to the Derveni Papyrus, its quotation from Herakleitos is a statement made up of two statements which, until its discovery, were considered to be separate & unrelated 'fragments' – if we referred to these statements as 'aphorisms', then their discovery as parts of a single sentence renders 'aphorism' as a misnomer & significantly alters the meanings that can be attached to them –

the conclusion would appear to be that I should use the term 'quotation' rather than either the time-honoured 'fragment' or the inappropriate 'aphorism' – but the idea of *The Quotations of Herakleitos* does not sit well as a title or a description of the subject – I will therefore adopt a term I have already used in passing, *The Sayings of Herakleitos* – and 'quotation' will be employed when referring to any instance of a Herakleitean text, even tho no term with a circumscribed scope will ever quite cope with a saying whose edges we are not altogether sure of –

♦ a number of scholars have used various techniques to highlight what they take as an authentic saying or quotation of Herakleitos, and a common one is the use of bold type for the 'authentic' bit while retaining regular type for what other words appear alongside it – often the fragment is actually a part of a longer sentence, and thought is given to whether or how such a part of a sentence can be so abstracted out as 'authentic' – mostly, the categories of statement do not go much beyond 'quotation', 'paraphrase', & some sort of other 'reference' – but Andrei Lebedev has looked at other possibilities for what kind of reference could be operative in any given case – accordingly, he makes the following list of options, which to my mind can serve as a very useful list of possibilities to bear in mind, but without employing them as guidelines for how to apply them to any particular fragment – Lebedev himself uses his list to identify the 'grades of authenticity' he finds in the texts – so, in his edition of the complete fragments he attaches these categorial labels to every part of every fragment – here is his list –

1. *Verbatim quotation* is like a 'brilliant uncirculated' coin (usually in Ionian dialect).
2. *Quotation* may contain some 'scratches', i.e. rewording, but still is very good.
3. *Paraphrase* – expresses Heraclitus' thought in later language. Some paraphrases (non-dogmatic) are hard to distinguish from grade (2).
4. *Doxography* – summary expositions of Heraclitus' thought in technical (dogmatic) terminology of later schools (Peripatetic, Stoic, Academic etc.). The use of doxography in this edition is limited. It should be edited separately.
5. *Reminiscence* – a brief allusion to Heraclitus' idea or saying *en passant*.
6. *Imitation or adaptation* – may contain authentic words and phrases, as well as genuine philosophical tenets of Heraclitus, but always requires independent parallels for confirmation.

7. *Reconstruction* – hypothetical reconstruction of the original on the basis of complementary versions or on the basis of ancient summary.[4]

this process, which I acknowledge is variously applied in varying degrees by some scholars, is a process I will not be undertaking here

◆ what does it mean, and what does it take, to 'authenticate' any group of words (any syntagm) as having been written by any particular person – for example, what if someone found a previously unknown letter by the poet Keats, in which was embedded a previously unrecorded sonnet, and which contained the poet's usual signature at the end – handwriting experts would soon determine that the letter's script was indeed that of Keats, and his attested and well-known signature would clinch the matter – the document is authentic Keats, independently of the content or the quality of the content in the document – however, if the document was not a letter, but a couple of sheets of paper, which contained the sonnet, but which did not contain a signature, epigraphic study could still find the hand-writing to be that of the poet, and the document with its previously unknown sonnet would join the body of Keats's writings – this is authentication by document, by the actual paper with the actual handwriting on it – what's more, the evidence in favour of its authentication is *verifiable*, it can be checked again and again – this form of authentication would be required, sine qua non, of any hand-written document purporting or purported to be the work of a particular person, poet, novelist, philosopher, anyone at all –

however, the sayings of Herakleitos cannot even be a candidate for this form of authentication – there are simply no extant handwritten documents that are presented to us as penned by Herakleitos – in that case, what can 'authentic' mean, what could it possibly look like – as it happens, we know a great deal about what exactly it looks like, and the example of Lebedev above is an especially elaborate version of it – here, the appeal is not to verifiable documentary evidence (because there isn't any), but to the authority of the commentator – the process seems to work like this : a person works in the field for many years, decades, and comes to an understanding of the philosopher's language and his themes – on the basis of this understanding, the scholar/commentator assesses the fragments for consistency of thought & language, and this allows him/her to apply this understanding to any fragment to see if the fragment fits conformably into the wider text – I do not say that this process is invalid in any way, nor do I suggest that, without verifiable documentary evidence, any conclusion must be incorrect – the prospect of further discoveries can certainly confirm any such conclusion – but all such conclusions are *provisional* until concrete evidence shows up to confirm or reject the initial postulate – even so, I myself am unable to use the term 'authentic' beyond questions of verifiable documentary evidence

◆    the question of how to spell the name of our preSocratic philosopher is a bit vexed in the literature – 'Heraclitus' is a common usage, but 'Heracleitus' is also known, and 'Herakleitos' is also used – major figure in this field Hermann Diels titled his edition *Herakleitos von Ephesos* – two Greek words and one German – the Center for Hellenic Studies (Harvard) is clear that Latin transliterations that have become conventional will be used in all their publications – so there's a fair amount of discrimination being exercised and decisions between options being made – yet, some Greek names are the same across both options : Parmenides e.g. is both an exact transliteration from the Greek and a conventional Latin usage at the same time – the same goes for Simonides and Sappho, but one finds both Archilochos and Archilochus, Ibykos and Ibycus (and Ibykus in German) in the literature – so we can choose from the uses that have been current in the last 100 years or so – in the editions I own, Kahn, Marcovich, Robinson, Heidegger, Bywater, Kirk, use 'Heraclitus', Jones in his Loeb edition & Kathleen Freeman in her *Ancilla* use 'Heracleitus', and Diels, Davenport, Bringhurst, use 'Herakleitos', which latter is my personal preference, partly because I prefer the challenge of pronouncing the name with the accent on the second syllable as Herákleitos (sounding more Greek) rather than on the third syllable as Heraclítus (sounding more English) – whether this is because I prefer originals to copies, or I like the sound it makes, is anybody's guess – on another hand, I will quote everyone's usage, of whatever spelling, accurately from their texts

◆    for a beginner studying ancient Greek, it can be confusing to note that, in the textbooks, a distinction is made between direct speech and indirect speech, which is generally signalled by the separate terms λέγει ('he said' in direct speech) and φησί ('he said' in indirect speech) – nowhere in the textbooks is it proposed that the terms are interchangeable – thruout

Herakleitos there are some twenty appearances of derivatives of φημί, and only two appearances of forms of λέγω, one in D46 and another in D75 – yet I find that most scholarly decisions about which is meant in any instance are made by referring to the *meaning* of the sayings in which they occur, rather than to their *grammar* – normally, direct speech amounts to quotation, and indirect speech amounts to paraphrase – is this how the Greek is to be read – I honestly do not know – but if one did go by the textbooks, then something like 95% of 'he said' phrases are paraphrase, and barely 5% are direct quotation – yet the scholarly assessment of which phrases are genuine quotations amounts to a much higher percentage than 5% and in some cases up to 75% or more – to go by the grammar, one would have to say that nearly all the sayings of Herakleitos we have are paraphrase, and not quotation

♦  lack of documentary evidence is sometimes filled with myths or stories or positions which allow commentary to reach where it may not have otherwise done so – one such story is that Herakleitos wrote material that was opaque or dark or obscure or paradoxical or enigmatic or 'something puzzling' that has permitted later 'readers' to embark on investigations based, not on documentary evidence, but on statements *about* the philosopher, made by people hundreds of years after the poor man's death, and which, in my reading, have never been questioned – I have to say I am tempted to try to be the one who illuminated the Philosopher's Dark, drew back the Curtains of Obscurity of his 'writings' that actually no one living has ever seen – but alas, I can only offer the notion that, while some of the Herakleitos sayings are a bit complicated, it's impossible for me to accept that current commentators are bamboozled by them – they seem no more difficult to understand than any other statements made by a sophisticated thinker, and essay after essay these days is often very clear about what it thinks Herakleitos was actually saying, or meant to say if the text is corrupted in any way – nor do the extant sayings suggest to me that Herakleitos was trying to befuddle anyone at all – he spoke plainly about matters that are not always plain – as a non-mainstream poet, I am only too well aware how readily mainstream poets and scholars can believe one is deliberately fudging matters, but I have always tended to write in plain English, even tho I have often played with syntax in unexpected ways – the view I take here is that Herakleitos is not, nor did he wish to be, 'Obscure', and I will read the sayings as if they do not contain linguistic or mystical booby-traps for the unwary – this approach is not about solving age-old mysteries, but about trying to read an old text which is spread before us in less than ideal condition

in what follows, I use a few abbreviations for sources –

Cunliffe = Richard John Cunliffe, *A Lexicon of the Homeric Dialect*,
  University of Oklahoma Press 1963.
LSA = the *Abridged version of Liddell and Scott's Greek-English Lexicon*,
  Oxford University Press 1990.
LSJ = *A Greek-English Lexicon*, Henry George Liddell and Robert Scott,
  revised by Henry Stuart Jones, Oxford University Press,
  ninth edition 1958.
SOED = *Shorter Oxford English Dictionary*, Oxford University Press 1959.

♦ I am struggling with whether addressing Jacques Derrida's notes on 'logo-centrism' is useful, relevant, or necessary –

I am also troubled by Juliet Fleming's *Cultural Graphology* in which she describes her topic as 'writing conceived without the guardrails of the book',[1] –

*'guardrails* of the book' – a book has no physical guardrails – a guardrail is a border or boundary designed to protect, prevent, or deter access to danger-ous or off-limit areas – if the word has no literal force, then is it merely a rhetorical device, a mere metaphor, with no referent in the outer world – ordinarily, if a work is published, then no reading is off-limits –

what if Fleming wrote 'writing conceived without the book' – and if so, why only the book – why not every other text-bearing 'substrate' – is it possible to conceive writing without a substrate – imagine that every single surface with writing on it disappeared from the earth today, actually, just this second – now – no clocks, no business cards, no passports, no product labels, no laws of the land, no literature, no bank statements, no street signs, no email, no books, pamphlets or magazines, no posters, no bus, train or tram timetables, no correspondence, etc – in that event, how could *any* notion of 'writing' be conceived at all

♦ Plato wrote he was averse to writing – Parmenides wrote his writing in verse – Whitehead wrote that the history of philosophy is a series of foot-notes to Plato – Gadamer wrote that our 'knowledge' of the preSocratics is inevitably filtered thru Plato – Derrida spent his life writing one of those footnotes –

but as far as eye can see, ancient Greek λόγος made no distinction between talking and writing – λόγος was any account whatever – yet the only means we have of 'knowing' about λόγος or any other term from ancient Greece is thru *writing* –

the Derridean problem with 'logocentrism' is a problem with Plato's reading and the christian reading – how far will I get having nothing to do with either –

there is no primary source in this discussion – one can only join in the chatter that has been in progress for centuries – or, one can exercise one's own set of prejudices among other appropriations of the preSocratics, and see what, if anything, is revealed –

Gadamer bared the difficulties and the task ahead – Heidegger dug into the ancient language instead of the history of its misuse – Wittgenstein figured that ordinary language was the place to work out what it is we can say – Jane Ellen Harrison kept the knot tied between the thing said and the thing done

◆   to what extent, when many authors propose 'writing', is it actually 'print' that they are referring to – is Derrida's 'Hegel', for instance, the Hegel of Hegel's autograph, or the printed and published Hegel – and if the latter, what edition – is it an edition in which Hegel's words and the words of his transcribing students are conflated without any typographical markers differentiating them –

to elide printing from the record of 'writing' is to elide the work of others – in the case of Hegel, 'others' means both the writings of several other persons, & the labours of several other people in the printing works that made the books – I wonder what a Derridean might make of Randall McLeod's essay on the first edition of *Orlando Furioso*, in which he shows that there can be many differences in the printed text *within a single edition*[2] –

even if one countered by saying that, before 1440, all texts were handwritten, one would have to counter *that* by pointing out that most such texts were not penned by their authors, but by copyists, many of whom would have been making copies of copies, and thus at a similar remove from 'writing' as is 'printing' –

how will we manage to distinguish 'writing' that is the act of composition, and 'writing' that is the act of copying, and 'writing' that is in fact 'printing' – yet composition and copying and printing have long had complex interrelations with each other, even in the apparently harmless activity of an author who changes their text (as an act of composition) by writing by hand on a printed (and thus copied) proof – and indeed, the very act of typesetting, an act of copying par excellence one might have thought, is also known as 'composition'

♦   is it possible that the entire Derridean project (which includes his followers, students, translators, publishers, & so on) has not examined the human act of writing by hand – even as it seeks to value 'writing' above 'speech' – is it fair to ask : why the binary – is not this an example of the very 'paradigm' that Barthes, for example, ardently sought to outwit[3] –

writing there is – printing there is – speaking there is – copying there is – where's the hierarchy – who put it there – to what purpose –

is the push to devalue speech a push to dehumanise language, to get all tone, gesture, emphasis, the body, out of the equation – & is that the same impulse as that of 'writing' that strips all physical and historical information away from the exigencies of print, even when it is 'print' that Derrida is engaging with, that is, reading, under the heading of 'writing' – and is this part of the almost pure privilege of the educated middle class : to elide the generally working class tradespeople and their labours from the very objects and connectivity with matter that those tradespeople and other workers provide as the ineluctable possibility of reading in the first place – is it part of a pattern I have noticed over many years, in which educated people tend to talk of 'concepts' rather than of 'objects' – and how many of those have we seen claiming that their 'concepts' have 'material reality' while the objects themselves and the people who make them are nowhere to be seen –

there is of course one place where all specific speech and print usages are held in abeyance – held at bay – the dictionary – philosophy as lexicality – where the multiple and contradictory meanings of each word are listed without preference or bias or valuing one meaning over another – is this the Neutral so sought in Barthes, Blanchot, et al – the ἐποχή (epoche/suspension of judgement) of Sextus Empiricus, the ἀταραξία (ataraxia/absence of conflict) of Pyrrhonism and the Stoics, the Negative Capability of Keats, the Indeterminacy Principle of Heisenberg –

it is hard to avoid proposing that many authors in their references to other authors do so by referral and deferral to 'what is published' rather than what is written, printed, or copied

♦ here, I will quote bibliographer D F McKenzie –

It is quite remarkable, for example, how many texts [in the seventeenth century] imply some kind of direct address or dialogue. Milton's *Colasterion* is 'A reply to a nameless answer against…' *The Doctrine and Discipline of Divorce*. His epigraph – 'Answer a fool according to his folly, lest he be wise in his own conceit' – is a common imperative in the period. Wing lists 424 titles which begin in the form 'To the…'. 'Humble' addresses, desires, hints, petitions, propositions, remonstrances, representations, requests, supplications, and so on, account for another 327. Petitions, Proposals, and Propositions (ones which, not being 'humble,' are entered under 'P') number 317. 'His Majesty' answers, declares, or sends messages to another 30. Titles beginning with the words Animadversion, Answer, Antidote, Confutation, Dialogue (153 of this), Reflection, Refutation, Remarks, Reply, Response, Voice, and Vox, together number 604. 'A Letter' or 'Letters to' account for 802 items. The round total they make is at least 3,066 – a figure which excludes all separate-issues and reissues and reprintings, and (with the sole exception of His Majesty) every item (at least as many again) whose author is known to Wing and is therefore found under the author's name. This rapid interchange of highly topical texts, of short pamphlets with short lives, helped to break down the anxiety-provoking distinctions among speech, manuscript, and print.[4]

one might also ask how this triumvirate of speech-manuscript-print has been complicated by the internet, where many 'texts' are only available on YouTube as recorded speaking (all tone, gesture, emphasis, and body preserved, however temporarily), yet when those 'texts' are issued in print, they appear as writing: *sans* tone, *sans* gesture, *sans* emphasis, & *sans* body

*EXEUNT !!!!!!!*

♦ what is interesting to me about the phrases in Herakleitos which use the word λόγος and its cognates is that their translation often bears little relation to their commentary, so much of which moves from a fairly mundane bit of language to a Grand Theory of The Logos without obvious transition from one to the other – it seems instead more of a leap of imagination or hope or ideology than a well-marked track leading step by step from one place to the next –

in this apparent disconnect between translation and commentary, I have no interest in what Herakleitos *believed* – we can never know that, and I cannot figure out why anyone would want to try to know that – what I want to know is what the man *said* – if, via Gadamer, the mens auctoris is not a yardstick for the meaning of a work of art,[1] then the same (given that many 'works of art' are poems) is true of the meaning of a spoken or written sentence – I am also remembering Wittgenstein's expressed dislike of being asked 'What do I mean by that?...In most cases one might answer : 'Nothing at all – I say...'[2]

even so, it is admittedly hard for many people not to feel that the meaning of an utterance is somehow 'extra' to a sentence, that it is situated somehow alongside or behind or thru the sentence (the sentence as window) rather than *in* the sentence – or that a person's beliefs are situated beyond anything they might say or write or do rather than *within* what they say and write and do – but what if you don't think that, what if you think that the words create the meaning on the basis of trace elements active from all past utterances, writings, and printings of a given language or languages in dynamic relation with our everyday ability to continually invent new sentences in speech and writing that we have never made before – one might well feel that belief is not a means for creating meaning, but a mechanism to stop the creation of meaning in its tracks –

today I went for my (currently) daily walk without the walking frame, and settled into the Royce Hotel for chips and a drink and to stop my legs from wobbling, and opened a book I bought years ago and did not look at until today, *Heraclitus Seminar*, by Martin Heidegger and Eugen Fink,[3] – I had written the above paragraphs this morning, and this is what I found in the Translator's Foreword where he (Charles H Seibert) quotes an anonymous participant in the seminar saying 'More is said in the interpretation of the fragments than stands in them' – & Heidegger replies, 'The interpretation is hazardous' – this is 1966–1967, a couple of years or so after I bought Kathleen Freeman's *Ancilla to the Pre-Socratic Philosophers*[4] & Heidegger's *Being and Time*[5] – ('and *now* they tell me' –

but I do take heart from Heidegger – at the end of the second *Heraclitus Seminar* he speaks of philologist Karl Reinhardt (1886–1958), lamenting that Reinhardt's writings were not available to the seminar, and saying 'Reinhardt was no professional philosopher, but he could think and see',[6] – all, perhaps, one could ever wish for –

elsewhere in the *Seminar*, Eugen Fink refers to what he calls the 'philological problematic'[7] in the Greek of Herakleitos (which I take to refer to the problems associated with establishing a reliable text), remarks that it is 'important', but then bypasses it in order to move straight to 'the matter for thinking' of Herakleitos – he does not say *why* it is important, nor does he provide a location in the discussion where the philological problematic might need to be dealt with before moving on to something else – a bibliographer or textual critic might regard the 'philological problematic' to be the first order of business, but Fink wants to move directly to the 'matter for thought' for Herakleitos which he posits to be somewhere away from the language, even tho he says the language itself is problematic, and even tho his own readings are based on the language of the text taken *as a given* – this insistence is of course deeply Heideggerian also – but if the 'meaning' of Herakleitos is to be found *within* the text, then the philological problems need to be addressed, if not sorted out, beforehand, should they not

♦ except for Plato and Aristotle, there are approximately 25 sources, all secondary, for the fragments of Herakleitos – about 65% of those quotes come from five authors : Clement (26), Plutarch (19), Hippolytus (18), Stobaeus (10), and Diogenes Laertius (10) – of the rest, Origen has 4 quotes, four others have three quotes each, five others provide two quotes, and ten others have one quote apiece – not one is contemporary with Herakleitos himself – [death of Herakleitos c 475 BCE]

approximate dates for all 25 sources are, chronologically –

BCE
    Theophrastus, 371 – 287, 2 quotes
    Polybius, 208 – 125, 2 quotes
    Philodemus, 110 – 35, 1 quote
    Strabo, 64 – 24, 3 quotes
    Philo, 20 – 50 CE, 1 quote
    Arius Didymus, [1st century], 2 quotes
*6 authors, 11 quotes*

CE 1st century
    Plutarch, 46–120, 19 quotes
    Aetius, [1st or 2nd century], 1 quote
*2 authors, 20 quotes*

CE 2nd century
　　Marcus Aurelius, 121–180, 3 quotes
　　Clement of Alexandria, 150–215, 26 quotes
　　Hippolytus, 170–235, 18 quotes
　　Origen, 184–253, 4 quotes
　　Sextus Empiricus, 160–210, 3 quotes
　　Maximus of Tyre, [2nd century], 1 quote
*6 authors, 55 quotes*

CE 3rd century
　　Pseudo-Plutarch [various, 3rd century], 1 quote
　　Diogenes Laertius, [between 3rd and 4th centuries], 10 quotes
　　Plotinus, 204–270, 2 quotes
　　Porphyry, 234–305, 3 quotes
　　Iamblichus, 245–325, 1 quote
*5 authors, 17 quotes*

CE 4th century
　　Themistius, 317–390, 1 quote
*1 author, 1 quote*

CE 5th century
　　Stobaeus, [poss. 5th century], 10 quotes
　　Proclus, 412–485, 1 quote
　　Simplicius, 490–560, 1 quote
*3 authors, 12 quotes*

CE 12th century
　　Theodorus Prodromus, 1100–1165, 1 quote
　　John Tzetzes, 1100–1180, 2 quotes
*2 authors, 3 quotes*

five authors, Clement, Hippolytus, Plutarch, Stobaeus, Diogenes Laertius, account for 83 quotations out of 119, almost 70% of the total. Just over half of these (44 out of 83) are from the christian apologists Clement and Hippolytus, both of whom were either refuting Greek writings or incorporating them into their apologetics – of the rest, 19 provide 4 quotations or less –

an unkind view might suggest we are reliant for most of the sayings on the least reliable authors in terms of their motivations for quoting them – a kinder view would acknowledge that anyone who quotes another has their own motivations for doing so – a more neutral view might suggest that not all motivations are equal in terms of contextualising any given remark – whose quotations are we to accept – who, in any instance, is *we* –

however, there seems no reason to modify these figures because of their approximation – arguments about their accuracy, however valid on a micro-level, cannot deny the general and practical picture that they present on a macro-level

*A note on the sources*

T M Robinson has a very useful section on 'Sources and Authorities' for the sayings – he lists all the authors and their works from which the sayings are taken, and every source noted is a printed edition, the earliest of which is dated 1832 – of the 42 sources he gives, 20 are printed editions published or begun in the 19th century – 25 of these are printed in Germany, 2 in Paris, and 2 in Amsterdam – that leaves some 17 that are printed in English[1]

◆ thruout the footnotes in Kahn, he notes approximately 30 disputed or variant readings and emendations in the Greek texts – i.e. in 30 out of 125 quotations – almost none of the other editions record these differences systematically, altho Kirk and Raven do mention a few in passing – Davenport, Robinson and Haxton don't mention them at all – a few passages give more than one source, but some editors tend to quote one and omit the other – except, typically, Marcovich, who includes as many as he knows – even the number of Herakleitos sayings is variable – Marcovich has 125, Bywater 130, Davenport 124, Kahn 125, Robinson 129, Freeman 137, Braxton 130, Kirk 137, and Andrei Lebedev 160 –

Marcovich, however, is very detailed about variant readings by other scholars and tends to have firm views about their likelihood, validity, usefulness, and so on – not all his quotations are set with this same amount of such detail, but here is a sample of a Marcovich disagreement about what a particular word might be, is said to be, is changed to be, or is asserted to be, or ought to be, etc – it's Marcovich's Fragment 32 (Diels-Kranz 59) – and the word is γνάφων (gnaphone) –

I read γνάφων (from γνάφος 'carding comb'), because *this* word is explained by Hippolytus as τὸ ὄργανον… ἐν τῶι γναφείωι. Kirk 97 ff. defends the ms. reading γραφέων, with an unlikely rendering: 'Of letters [or, of writers] the way is straight and crooked' (in such a case one would rather expect γραφέως χειρὸς ὁδος or sim.). Guthrie 443 adopted Kirk's interpretation as 'convincing' ('The track of writing is straight and crooked').

But Kirk (Mullach & Hugh Lloyd-Jones ap. Kirk 101) have to presuppose the following changes: γραφ- (the original of Heraclitus) > γναφ- (the source of Hippolytus) > again γραφ- (cod. Paris). This is not likely to me. It is simpler to explain γραφέων P as a mistake or lapsus of the original γνάφων, since the same scribe Michael (who probably knew very well what γραφεύς and γραφεῖον meant, but was not quite sure about γνάφος and γναφεῖον) writes in the *next* line γραφείωι, in lieu of the sure γναφείωι.

As for the reading γραφέων from τὸ γράφος (adduced by Kirk 101 f. from *G. D. I.* Collitz, nrr. 1149, 7; 1151,19; 1156, 2; 3; 1157, 6; *I. G.*, V, 2, nr. 343, 18–19), my objections are: (1) the word is rather dialect (limited to the Peloponnese); (2) it means either 'inscription' itself or its contents [ = γράμματα, LSJ, III, 1–2], but not letters ΑΒΓΔ, which is the required meaning here.[1]

The following is Bywater/Patrick c.1888 –

Fragment 118 – Note 37 – The common reading is, δοκεόντων ὁ δοκιμώτατος γινώσκει φυλάσσειν, which makes nonsense. Schleiermacher proposes δοκέοντα ὁ δοκιμώτατος γινώσκειν φυλάσσειν. Schuster (p340) suggests, δοκεόντων ὁ δοκιμώτατον γίνεται γινώςσκει φυλάσσειν and fancies the allusion is to the poets, who from credible things accept that which is most credible. Bergk, followed by Pfleiderer, reads φλυάσσειν to talk nonsense. Bernays, followed by Bywater, reads πλάσσειν.[2]

Marcovich's note is about a single word, Bywater's is on whole phrases, & both relate to a single passage – add to these the number of other sayings with contested terms, and the entire Herakleitean text is shown to be unstable – but what we do not get is a facsimile or picture of the actual lettered/worded document that triggered these disputed responses in the first place – a discussion about the actual text is replaced by a discussion about the opinions of the scholars –

I cannot help thinking, however, that Marcovich's rejection of Kirk's 'The track of writing is straight and crooked' in favour of 'carding comb is straight etc' makes little sense in terms of the rest of the sayings of Herakleitos that we have – if 'writing' is seen as graving or incising letters, then they are vivid exemplars of the straight & crooked all-at-once of the things of the world, whereas the 'carding comb' provides no such example, where the contrast might better be straight and *curled,* and not the 'crooked' of the text – that most letters of the alphabet are both straight and *crooked at the same time* seems to me more to the point than that the carding comb simply has straight *here* and crooked (or more accurately 'curled') *there* – one can also wonder why Marcovich accepts Hippolytus so readily, as he himself does not say –

the number of assertive declarations about what Herakleitos is supposed to have meant or intended strikes me as disproportionate to the degree of instability inherent in the documents – I cannot deny the temptations of those who look to figure out what the sayings of any of the preSocratics might amount to as a coherent way of looking at the world, but I would much rather be acquainted with the actual texts & knowledge of the actual documents (papyrus or paper or skin) from which later printed compilations have been made –

it is hard not to come to the view that many of the editions, the translations, the commentaries on Herakleitos are trying to piece together a case for asserting Herakleitos's philosophical *system* – more about what he was supposed to have believed, and not what the text, taken as it is, might actually say – they are not looking for a *reliable* text, they are looking for a *credible* text with a sense of coherence that allows them to guess at what Herakleitos might have meant – & the presence, beside most of the translations, of sometimes considerable *commentary*, can only support this view – they seek to elaborate on what the texts ought, in their view, to *mean*, rather than let them speak no further than what the texts can be seen to *say* –

even so, my quest here is clearly its own kind of chimera, one which seems to value the copying error, slips of the pen, mis-readings, faulty typesettings, passages that, taken as they are, can lead only to nonsense, mis-spellings, foul copy etc in the hope of trying to read *what's there* instead of trying to establish what *ought to be there* – that my interest in what can be verified (the actual letters on the page) can only be at the cost of what can be understood – Marcovich e.g. lists variant readings by various scholars as part of the path to a reading of his own, and his decisions seek textual credibility based on his overall comprehension of Herakleitos's intended meanings –

of course, it looks as if I'm writing my own epitaph, trying to establish an unestablishable text – & I don't even have any information about the basic documents from which each of the readings are derived – I don't know what 'Clement of Alexandria' means as an original document – is it on paper or papyrus or skin, is it Clement's handwriting, or that of a copyist, or a later copied or printed 'Clement' for which the copy-text no longer exists – I just don't know, and the sources I have access to shed no light on it at all

◆  there are at least five reasons for being sceptical about the veracity and accuracy of 'quotations' from the 'writings' of Herakleitos –

1.  a quarter of the quotations are offered in religious contexts, none of which were contexts for Herakleitos himself – much current scholarship is clear about this, and Gadamer has expressly set about peeling some of the other purposes off the 'Herakleitos' that those early authors present – but the effort to read Herakleitos as a precursor to later religious & other 'systems' of belief began early, and continues still

2.  the first quotation is made just over one hundred years after the death of Herakleitos, and the last one around 1500 years after his death – there is thus a question of the time elapsed between 'author' and 'quotation' –

just as there is no logical argument against the proposition that any of those who quoted Herakleitos were correct, there is none against the proposition that any of the quoted material is incorrect – what any quotation has that the original does not have is the accumulated other writings in the culture between the time of composition and the time of quotation –

also across that time are issues, terms, language, cultural milieux, & movements that the original author can have had no knowledge of, and which inclines the later authors toward positions and questions that were not part of the understanding of the original author –

while it cannot be pinpointed, the passage of time has to operate as a mediating filter from which we have no guaranteed protection

3. what documents, exactly, were each of these later authors quoting and, where were they? –

whatever anyone knows about the handwritten MSS of Clement, Theophrastus, Hippolytus et al, none of that information reaches me thru the pages of the books I have – when I do look at the books I own, this is what I find out –

    a. from T M Robinson, every source noted is a printed modern edition – there's no account of the existence or whereabouts of any of the source material from which the printed editions were derived –

    b. from Guy Davenport, there is nothing at all about the edition(s) from which he may have translated –

    c. Kathleen Freeman's translations are from Diels, 5th edition –

    d. Brooks Haxton's work is based solely on modern printed editions, from Ingram Bywater in the 19th century onward –

    e. Charles Kahn cites both Diels and Marcovich in a Concordance, as well as a number of ancient writers and editors of recent editions – but there is no information on ancient MSS he may have consulted, nor in what libraries or museums they may be found –

    f. Kirk and Raven : similar to Kahn – wide-ranging scholarship and close reading, but no explicit reference to a single handwritten document –

    g. Marcovich, with a full critical apparatus (656 pages! & an attempt to provide the complete range of sources for anything that is, looks like, or may be, a source for a Herakleitean text –

of the printed sources in *the history of philosophy*, I learn of none before the 19th century, yet *the history of printing* is much more voluble – see K Sp Staikos, *Greek Philosophical Editions in the First Century of Printing*, European Cultural Centre of Delphi, Athens 2001; K Sp Staikos, *The Greek Editions of Aldus Manutius and his Greek Collaborators*, Oak Knoll Press, Delaware 2016; Nicholas Barker, *Aldus Manutius and the Development of Greek Script & Type in the Fifteenth Century*, 2nd ed, Fordham University Press, New York 1992; Robert G Babcock and Mark L Sosower, *Learning from the Greeks*, Beinecke Rare Book and Manuscript Library, New Haven, Connecticut 1994, among others –

none of the sources for any of the Herakleitos fragments was alive at or after the advent of printing in 1440–1445 – the latest of them lived in the 12th century when printing is about 300 years away –

none of the references to printed sources in my present day books are to any work printed before the 19th century, yet by then, printing had been active for some 300 years –

it suggests that there are no references, in the philosophical tradition, to any sources written or printed between the 12th century and the 19th century – 600 to 700 years of scriptural oblivion in which there was nevertheless a great deal of scribal activity in Greek, and a great deal of printing activity in Greek

4. nearly all printed Greek editions, from the 15th century onwards, present BCE writings in CE scripture – accents in Greek were introduced around the 2nd century BCE, and lower case letterforms some time after 800 CE – nearly all the 'ancient Greek' available to us in print is in a script with which none of the preSocratics (nor Plato, nor Aristotle, nor Theophrastus, nor Sextus, nor Plutarch, etc, would be familiar –

there may be more, but I know only three exceptions, where a preSocratic text is printed in capital letters: a selection from *The Greek Anthology*, compiled by Janus Lascaris, printed in Florence in 1494; a Guy Davenport *Herakleitos* (1990),[3] & a Robert Bringhurst *Parmenides* (2003),[4] – both published, not by a scholarly or university press, but by a private press printer, Peter Koch, in Berkeley, California – all modern editions are modalities of 'copying' which are thus at several removes from any possibility of 'original documents' – [a few weeks later, Peter Koch tells me of Alberto Tallone's edition of Parmenides in Paris, 1953, with the Greek text in sloped capitals, and with a French translation by Jean Zafiropulo[5]] –

for an abject lesson in how a text (in this case a single short poem) can be regularly edited and published by high-reputation publishers over several centuries, and *never reproduced accurately to an author's autograph that still exists*, see Randall M Leod's devastating account of the publishing history of George Herbert's 'Easter Wings' (first published in 1633) in *Crisis in editing: texts of the English Renaissance*, AMS Press, Inc, New York, 1994

24

5. is it the case that all editions since the late 19th century draw from Diels and/or Bywater, and not the documents from which both drew their 'fragments' – are they compilations derived from a compilation – when it comes to Herakleitos and the books I have, Kahn refers to Ingram Bywater (whose *Heraclitii Ephesii Reliquiae* is published 1877, 26 years before Diels in 1903), and so does Haxton, but Davenport, Robinson, and Kirk and Raven do not – Marcovich refers to Diels, and also to F Schleiermacher's 'First collection of the fragments' (which does not seem to be a book-title) of 1807 –

no one, however, in modern editions refers to the first printed edition of the preSocratics by Henri Estienne and J J Scaliger in Geneva, 1573 – except scholar-poet Robert Bringhurst in the *Afterword* to his translation of Parmenides – there he writes : 'Estienne had found ten fragments, roughly 70 lines, of Parmenides. Over the next decade or two, his friend Joseph Scaliger doubled the length of this corpus. ...But no one ever published Scaliger's enlarged collection of the fragments.' In a note to this Afterword, Bringhurst says that Scaliger's expanded ms still exists in the library of the University of Leiden. I am wondering if the text of Herakleitos collected by Scaliger is in the same ms Scal. 25 at Leiden. On the question of the role of the Estienne book in philosophic history, Bringhurst is clear : it 'appears to be one of the key texts in the history of European thought. At the time it had little discernible impact outside a tiny circle of friends'

♦ POESIS PHILOSOPHICA, Geneva, Stephanus, 1573. 8vo. The very
rare first edition of the first anthology of pre-Socratic thought. This mile-
stone publication, edited by Henri Estienne and with Latin notes by J.J.Scali-
ger, constitutes the first printing of the surviving fragments of Pre-Socratics
– the originators of Greek philosophy – and the first work devoted solely to
Pre-Socratic thought. Up until it was published, the modern world had only
known the Pre-Socratics indirectly, e.g. through references in Plato or Aristotle. "Modern interest in early Greek philosophy can be traced back to 1573,
when Henri Estienne (better known under his Latinized name Stephanus)
collected a number of Presocratic fragments in "Poesis philosophica"." (Gian-
nis Stamatellos: "Introduction to Presocratics", p. 7). Within the history of
modern philosophy and modern thought in general, the importance of the
present work can hardly be over-estimated. Containing for the first time on
their own the fragments of the pre-Socratic philosophers: Heraclitus, Parmenides, Pythagoras, Empedocles, Xenophanes, Cleanthes, Democrates,
etc., this foundational work is "A VOLUME OF MAJOR IMPORTANCE
TO THE HISTORY OF WESTERN THOUGHT, which rightly belongs on
the same shelf with the first editions of Plato and Aristotle". (Schreiber).
As Plato and Aristotle form the foundation for Western thought, so the Pre-
Socratics form the foundation for Plato and Aristotle. "The first philsophers
paved the way for the work of Plato and Aristotle – and hence for the whole
of Western thought. Aristotle said that philosophy begins with wonder,
and the first Western philosophers developed theories of the world which
express simultaneously their sense of wonder and their intuition that the
world should be comprehensible. But their enterprise was by no means
limited to this proto-scientific task. Through, for instance, Heraclitus'
enigmatic sayings, the poetry of Parmenides and Empedocles, and Zeno's

paradoxes, the Western world was introduced to metaphysics, rationalist theology, ethics, and logic, by thinkers who often seem to be mystics or shamans as much as philosophers or scientists in the modern mould." (Robin Waterfield (edt. and transl.): "The First Philosophers: The Presocratics and Sophists", introduction). This first edition of the foundational fragments, the fragments of the works upon which Western thought rests, gave to the modern world the opportunity of studying Pre-Socratic thought as such. Henri Estienne's 1573 publication of "Poesis philosophica", the first anthology of pre-Socratic thought. "Diverse fragments of Heraclitus were found in the works of Plato, Aristotle, Plutarch, Sextus Empiricus, Stobaeus, among others, but were generally presented there in other philosophical contexts. Estienne's anthology, which contained some 40 Heraclitean fragments, or about one third of what we have today, allowed Heraclitus to be read on his own and provided the opportunity to grasp the overall philosophy of the master of the oracular epigram." (Jerry C. Nash's review of Francoise Joukovsky's "Le Fue et le Fleuve: Héraclite et la Renaissance francaise"). With this publication came the revival of Pre-Socratic thought and the birth of Pre-Socratic scholarship. Being now available to Renaissance thinkers, the Pre-Socratics could now be studied as they deserved, and this seminal publication is responsible for the direction that much modern philosophy was to take, for centuries to come. Only in the 19th century did a new edition of the Pre-Socratic fragments compete with Estienne's great anthology. "How did you study the Presocratics in the Renaissance? In 1567 Élie Vinet did so by commenting on a late Latin text, itself adapted from Greek sources: the "De die natali" of Censorinus. In less than a decade Henri Estienne and Joseph Scaliger did so in a far more original and systematic way, by collecting and analyzing fragments quoted by Clement of Alexandria, Simplicius and Sextus Empiricus." ("Estienne and Scaliger 1573"). (C.B. Schmitt, The Cambridge History of Renaissance Philosophy, p.767). With this publication, for the first time since Plato and Aristotle, the Pre-Socratics came to once again directly influence Western thought. For instance, "[Francis] Bacon could have found an important precedent in Henri Estienne's "Poesis philosophica" (1573), a collection of the fragments of pre-Socratic poets that included Empedocles. After Estienne's book appeared in print, the idea of assembling pre-Socratic wisdom became more common among scholarly men of Bacon's generation." (G Passannante, "The Lucretian Renaissance", p. 145). Adams P-1682; Schreiber 142.

◆   I want to look at the advertisement above, and see whether its claims have consequences and/or resonances in what I can find out about modern editions, from the 19th century to the present

*the first claim* : 'edited by Henri Estienne and with Latin notes by J.J. Scaliger, constitutes the first printing of the surviving fragments of Pre-Socratics – the originators of Greek philosophy – and the first work devoted solely to Pre-Socratic thought. Up until it was published, the modern world had only known the Pre-Socratics indirectly, e.g. through references in Plato or Aristotle'. – but even after the book was printed, the 'modern world' still only knows the preSocratics indirectly, thru Plato and Aristotle and the other later commentators (Clement, Hippolytus, etc –

*the second claim* : 'Within the history of modern philosophy and modern thought in general, the importance of the present work can hardly be overestimated.' – while it's clear that it is an important work in *the history of the printing of the preSocratics*, I can find no evidence in the modern editions I have of any influence of this book on modern studies, editions, or translations, not even in Marcovich, whose range of reference to other works is far greater than those of the other editions –

*another point* : the preSocratic texts in this volume are all in Greek, which one might have thought was a primary reason for going to it and comparing this edition with other later editions –

what attracts me in the above advertisement for the 1573 edition is that it reads like a perfectly respectable account – it is considered, informed, detailed, learned about the history of philosophical studies, and on the face of it, perfectly convincing – and yet it seems to be dead wrong as a story of this edition's influence and significance in the field –

furthermore, it is not only that works survived in bits & pieces rather than in full texts, but there were many authors whose words have not survived in any fashion or amount – in Kathleen Freeman's *Ancilla*, she records a total of about 55 preSocratic philosophers or authors of whom 'no writings survive', and two of whom she says are 'Heracleitians' – of course, commentary tends to attach itself to texts that do exist, but how do we understand what happens to our sense of things when we allow ourselves to think of the loss of the texts that have not survived, if for no other reason than many of those texts and partial texts will have nevertheless formed part of the context in which other texts were written and how they were read and received –

how many of those lost words might have been describable in Anne Carson's wonderful phrase, 'a fragment of unexhausted time' – of course, we cannot know, but can we not allow for the possibility to act as a brake on our capacity to assume unwarranted certainties of text, context, and interpretation

Socrates is said to have commented, on being shown a work by Herakleitos, that to understand some of it one would have to be like 'a diver from Delos' to plumb its depths – Delos of course is one of the islands in the Cyclades, & the divers were pearl fishers – to read him, we have to 'inquire into many things' as Herakleitos is himself said to have said – for that is all we have : quotes, paraphrases & speculations by others all centred around what Her-akleitos is said to have said – the word δῆλος (dēlos) in ancient Greek meant 'manifest', 'visible', 'apparent', 'appearance', and its associated adverb δῆλον (dēlon) meant 'clearly' or 'plainly' – a pub-test might suggest that there is something fishy about claims that Herakleitos said this, or meant that, or believed these things – all I can say about my own deep interest in this ancient curmudgeon is that, as others before me, there is something about those flashing, half-clear bits of language that strike, quick as lightning, to a heart of things which also has multiple hearts – our share in 'the general implication' (A N Whitehead) – maybe the thing to do is to simply remain as one of the fish, one of the divers, whose job is not to bring anything to the surface (the apparent task of professional philosophers) but to better learn how to navigate the dark, & to describe its lineaments, to show how to think the dark and think *in* the dark – Herakleitos is to *remain* the Dark One, or the Obscure, however deeply we dive and delve into his apparent yet never to be taken for granted 'writings', which are never his

what seems to me to be emerging now is a disjunct between the efforts of
'the history of philosophy' & the efforts of 'the history of bibliography and
textual criticism' – the former seeks an answer to the question, What did
Herakleitos mean? (among the textual variants & differing historical inter-
pretations), and the latter seeks an answer to the question, What are the
marks on the MSS? (independent of speculations about what the marks on
the MSS *ought* to be) –

and here's another thing, from Rochelle Altman from her *Absent Voices:
The Story of Writing Systems in the West*[1] – Altman makes a distinction
between 'semantic-based writing systems' and 'phonetic-based writing
systems', and it's worth quoting a fair chunk about how she understands
the difference –

> Modern writing systems are semantic-based; their purpose is to con-
> vey data. In antiquity, 'the voice of authority' was not a metaphor to
> be tossed about by literary critics, but a concrete, visible reality. For
> more than four of the five thousand year history of writing, writing
> systems were phonetic-based; their purpose was to record the voice
> of absent authority – be it poet, priest, or king.
>
> Accustomed as we are to semantic-based systems, we incorrectly
> are taught to believe that we can use any script or type style we wish,
> and it will make no difference. We erroneously assume that the size,
> format, and script of a text is unimportant and irrelevant.
>
> But semantic-based or no, our modern writing systems use hier-
> archies of scripts, sizes, and formats to identify a text. We automati-

cally associate size with authority, format with content, and script with a language or a people. ...Semantic-based writing systems are rooted in data transmission; their primary purpose is to convey information. Phonetic-based writing systems are rooted in oral communication; their primary purpose is to convey a record of a 'text' as spoken – with all those factors that serve to identify that specific text as itself and no other (an entity)...Each component of these systems holds meaning. ...The ancient texts contain more than words. Every part of a text written in these systems holds meaning. ...Our modern use of the components of a writing system is completely unconscious; we use them, but we do not know *why* we use them. In antiquity, people knew the meaning and relevance of every component of a writing system.

what if, then, Herakleitos wrote at least some of the time in a phonetic-based writing system, and the entire modern history of philosophy treats him as if he wrote only in a semantic-based writing system – one way of looking at it is that, in a semantic-base, the phrase 'in other words' can have genuine operative power – in a phonetic-base, the phrase 'in other words' can have no place at all – yet another way is to say that semantic-based statements are paraphraseable, and phonetic-based statements are not, as they rely on the exact words in their sequence and arrangement (like poems) for their total communicative effect – for a supportive view, this is Eva Brann's take : '... Heraclitus, an engaged solitary, an inward-turned observer of the world, inventor of the first of philosophical genres, the thought-compacted aphorism, *prose that could contend with poetry.*'[2] [my italics], tho whether Herakleitos wrote 'poetry' or 'prose' as we understand these terms, is another question –

Wittgenstein : *Do not forget that a poem, even though it is composed in the language of information, is not used in the language-game of giving information.*[3]

most translators seem to want their efforts to result in a good English sentence – but what if that is precisely *not* the thing to do – the history of the American Language Poets shows that mellifluous English was not the vehicle for the writing they wanted to do – many used phrase constructions that worked against the prospect of a good English, or good American, sentence – perhaps the rugged shape of an early Greek sentence should be mirrored in a more ruggedly shaped English one – if Herakleitos was not using the standard poetic structures of his day (around the hexameter), & nor did he write in the sort of prose practiced by Herodotus, Thucydides, or Plato, then why would a translator insist on choosing one of those modes for their English – for me, the notes by Havelock are determinative – 'the first heave' as Ezra Pound had it, was of the pentameter, but the hexameter was broken by Herakleitos long before –

in any case, I think it's fair to propose that modern editors/translators of the preSocratics are concerned to write good English sentences which tend to be semantic-based, where some of the texts they are translating tend to be primarily phonetic-based – the translations of Guy Davenport, however, are clearer that there is something like poetry in their originals (Eva Brann's 'prose that could contend with poetry' – and that is reflected in the sort of English his translations achieve – poetry rather than prose, altho less like the Language Poets than the Open Form poets of the 1960s –

if we treat phonetic-based sayings as semantic-based ones, then we have no compunction about getting over our difficulties by saying 'in other words…' which, for a phonetic-based utterance, we should not do – even tho all translation is in some sense a matter of 'in other words', one language to another

◆   in an example from another language, Chinese, here is a shift from a phonetic-based gloss to a full semantic-based translation from Lewis Hyde, working on the famous Chinese Oxherding pictures – Hyde[4] has 3 stages, first, the gloss, then a 'literal' translation, then a full, final, English version –

*Without bounds*
>            *stirring*            *grasses*
>            *leaving,*            *tracking*            *down*

*Waters broad*
>            *mountains*            *distant*
>            *road*            *more*            *obscure*

*Strength exhausted*
>            *spirit*            *weary*
>            *no*            *place*            *to-hunt*

*But hearing*

        *sweetgum*      *trees*
        *evening*      *cicada*      *song*

now the first 'literal' translation –

*Searching, pushing through endless underbrush.*
*Wide waters, distant mountains, darkening path.*
*Strength exhausted, spirit weary, no hint of where to hunt.*
*Just hear the evening cicada sing in the sweetgum grove.*

and then the full English translation –

*Alone in the deep woods, despairing in the jungle, searching in darkness!*
*Flood-swollen rivers, mountains beyond mountains*
    *the trail endless and unchanging.*
*Bone-tired, heart-weary, the whole thing seems hopeless.*
*No sound but the evening cicadas singing in a grove of maple trees.*

the trick here is to move from the initial gloss to the final English version, and see what has been done to achieve a semantic-based work – it seems undeniable that much compact energy is dispersed when the text moves from gloss to English, and the semantic-based translation contains meanings that the phonetic-based gloss does not have – *the whole thing seems hopeless* in the full version is a long way from *no   place   to-hunt* of the gloss – especially when 'to-hunt' is apt to the search for an elusive quarry –

◆   another example is in Robert Bringhurst's translation of Parmenides – here is his gloss of the first 7 lines [5] –

ΙΓΓΟΙ ΤΑΙ ΜΕ ΦΕΡΟΥΣΙΝ ΟΣΟΝ Τ.ΕΓΙ ⊕ΥΜΟΣ ΙΚΑΝΟΙ

    Mares that me carrying as-far-as over heart reach

ΓΕΜΓΟΝ ΕΓΕΙ Μ.ΕΣ ΟΔΟΝ ΒΕΣΑΝ ΓΟΛΥ⊕ΒΜΟΝ ΑΓΥΣΑΙ

    conveyed when me in road went many-voiced leading

ΔΑΙΜΟΝΟΣ Β ΚΑΤΑ ΓΑΝΤ.ΑΝΤΒΝ ΦΕΡΕΙ ΕΙΔΟΤΑ ΦΩΤΑ

    holy-one's that along all straight carry knowing mortal

ΤΘΙ ΦΕΡΟΜΘΝ ΤΘΙ ΓΑΡ ΜΕ ΠΟΛΥΦΡΑΣΤΟΙ ΦΕΡΟΝ ΙΠΠΟΙ

On-it carried on-it for me much-spoken carried horses

ΑΡΜΑ ΤΙΤΑΙΝΟΥΣΑΙ ΚΟΥΡΑΙ Δ.ΟΔΟΝ ΒΓΕΜΟΝΕΥΟΝ

chariot straining girls so road steered

ΑΞΩΝ Δ.ΕΝ ΧΝΟΙΘΙΣΙΝ ΙΕΙ ΣΥΡΙΓΓΟΣ ΑΥΤΘΝ

Axle so in bushings squeal socket's cry

ΑΙΘΟΜΕΝΟΣ ΔΟΙΟΙΣ ΓΑΡ ΕΓΕΙΓΕΤΟ ΔΙΝΩΤΟΙΣΙΝ

burning both for hard-pressed whirling taking...

and, from the same publication, here is the final translation –

> *Racehorses take me. They stretch me. They pulled me*
> *as far as the heart can bear when they ran with me*
> *straight up the track that passes through everyone's voices.*
> *They carried me, a non-god, open-eyed*
> *the whole way, down a holy being's trail.*
>
> *That's where I was. I was pulled there*
> *by thoroughbred mares. Girls*
> *worked the reins to keep the wagon,*
> *going flat out, from flying off the road.*
> *The wheel-bushings screamed. The wagon-axle twirled*
> *like a double-ended fire-drill...*

as a further experiment, let me take Bringhurst's gloss, & see what happens when I try to 'translate' that into an English which seeks to retain some of the ruggedness of the Greek in a ruggedness of English – (I'm taking Bringhurst's gloss as a given, not arguing here over such questions as whether ἵπποι should be 'mares' or 'racehorses' or 'horses') – Bringhurst in italics –

*Mares that me carrying as-far-as over heart reach*

    Mares carrying me beyond heart-reach

*conveyed when me in road went many-voiced leading*

    conveyed when in the many-voiced road

*holy-one's that along all straight carry knowing mortal*
 the daimon leading-carrying this knowing mortal

*On-it carried on-it for me much spoken carried horses*
 carried much-spoken me, carried horses –

*chariot straining girls so road steered*
 girls so road-steered the chariot

*Axle so in bushings squeal socket's cry*
 that axle squeal, socket's cry

*burning both for hardpressed whirling taking…*
 both burning in hardpressed whirl, taking

I have kept 'daimon' (which, over decades, has achieved some currency in English) – but still, it seems to retain a sense of poetry, a sense of the rugged nature of the Greek, and enough readable English not to expand on the language to conform to notions of a good English sentence – Bringhurst takes seriously that Parmenides's work is a poem written in archaic hexameters, 'and contains numerous thematic and stylistic echoes of the *Odyssey*' as David Gallop writes in the Introduction to *his* translation of Parmenides – it's a poem as much as a treatise, and Bringhurst's translation reflects that very well – yet there is something in the compacted Greek inflections, and the word-order within the lines that does not follow the conventions of English word-order, that I wish to retain, perhaps wilfully as the often unconventional poet I have been for fifty years, in the torque and tension in the impossibility of rendering in one language what has been first tendered in another –

even so, it's not easy to come up with ways of understanding this weird situation of having and not-having a reliable text which all of us can equally respect and suspect – again, Heidegger is a bright guide : '…one can easily criticize any translation, but can only rarely replace it with a 'better' one'.[6]

♦   when it comes to how this phonetic/semantic distinction might work in
relation to the difference between a bibliographic approach to an ancient text
& a philosophic approach to the 'same' text – it seems that the bibliographic
approach is happy to ask : What do the actual marks on the page say? and the
philosophic approach is happy to ask : What is it that these marks on the page
mean? The bibliographic approach is phonetic, the philosophic approach
semantic –

what and where are the sources for Herakleitos – we know which writers
were said to have quoted him, but what and where are the documents from
which we derive this information – for my purposes, these documents cannot be printed ones, and therefore must date well before 1440, and must be
handwritten on papyrus, paper, or skin – do any of them exist, & if so, where
are they – and if they do exist, are they genuinely Clement or Hippolytus or
are they too copies of copies etc – I can't help thinking these may be unan-
swerable questions, but I do not *know* that they are in my present state of
general ignorance, and the books I have access to do not tell me otherwise

♦   today [several days after writing the above] a small penny hath dropped
– the reason only printed sources are referred to is that scholars spread all
over Europe and the Americas cannot realistically have access to papyrus etc
documents – the printed sources are paradoxically both 'secondary' and 'primary' for nearly all scholars in the field and are thus essential in informing
anyone about the texts and their 'philological problematic', as Eugen Fink
put it – it means that only a tiny number of individuals can possibly have
access to the written documents upon which all printed documents are ulti-

mately based – those printed sources therefore are only likely to have their texts questioned when someone re-edits the written sources for a new printed edition – could one say that the printed documents are better referred to as 'resources', while the written documents remain as the 'sources' –

so for me, there are only the printed resources to turn to, & many of those are not written in English, which leaves me with English resources at a number of removes from any possible ur-text, or even the earliest printed resources – I have to make do with the documents I can have access to and can read

♦ all the material I can find on making and reading a Critical Apparatus depends on a knowledge of Latin – I don't have the time or the overall need to do that –

when W H S Jones in his Loeb edition of Herakleitos writes 'See…Philo, Rev. Div. Her. 43' (p455) he seems perfectly comfortable with the abbreviation, and has no apparent need to tell his reader what kind of text it is, nor where it is housed – as a fairly typical remark in the literature, it's clear that the scholars are not writing for me or any other general reader –

for evidence of the role of personal opinion in the literature, here's Jones again : 'the rule I have tried to follow is to record only those readings that are intrinsically interesting and those that seriously affect the meaning' (pvii) – well, what about those readings that *subtly* affect the meaning, are there none of those, and if there are, are they irrelevant – & what about 'intrinsically interesting', is that simply what interests that particular writer, is what he finds so what I would find so, and the next door neighbour would also and 'naturally' find so – when the scholars say 'See' they are addressing other scholars, seem always to be referring to a printed edition rather than a written one –

Jones also poses then a fine statement of one of the problems in reading an ancient author, then undermines it two pages later : 'It is both confusing and depressing to read the treatises of Lasalle, Teichmüller and Pfleiderer, to see how the most opposite and inconsistent conclusions can be drawn by learned and intelligent men from exactly the same evidence' (p453) – and then : 'What Heracleitus really meant, and should have said, is that…' (p455) –

the references in Liddell & Scott seem to be to printed sources – even so, it also appears that one needs highly specialised training to be able to read their critical apparatus and other references in the full LSJ –

telling note in the LSA : '…especial care has been taken to explain all words contained in the New Testament' (October 1871)

♦   a recent bilingual edition of Herakleitos, privately printed in England, and without stating the name of the editor/translator (one reason why I decided not to buy it), has translated λόγος as 'Word' with a capital W – this is not a 'translation', but a *quotation* from the gospel of John in the King James version of the New Testament : 'In the beginning was the Word', published in 1611, nearly 2000 years after Herakleitos is supposed to have written 'λόγος' – and John (however many of him there were) is said to have written his 'λόγος' c.90–100 CE, almost 600 years after Herakleitos wrote his – the prospect that these three 'λόγοι' can be read as having the same meaning when they appear in such radically different contexts many hundreds of years apart needs, at the least, some considerable justification – in any case, the later versions are a long way from the 'sober thinking' Herakleitos himself advises – after all, how many of us have had conversations with contemporary others in which some terms, like 'sublime', 'transcendental', 'postmodernity', 'spiritual', among heaps of others, have been in dispute – the minute words are, as it were, lifted from the day to day language of the time & turned into more or less technical terms for use by specifically-educated experts in an intellectual field, then there is no longer any universal agreement about their use in the wider community – another way in which this is done is where someone from one language/culture takes a word from another language/culture and uses it for different purposes altogether – so one could ask : At what point did the λόγος of Herakleitos get lifted from an ordinary word in the conversation of a people & made into a technical term for specialist use – for some writers it is within the time and writing of Herakleitos himself – but what happens to one's understanding of Herakleitos if one takes it that some or all of those transformations from everyday use to specialist use came after his death, that the transformations were made by other people with reasons of their own, and which were not those of Herakleitos – my over-riding authority here, *of course*, is Humpty Dumpty in *Through the Looking Glass and what Alice found there*,[1] & here is the appo-

site bit of conversation twixt Humpty Dumpty & Alice about the meaning
of words :

> 'When *I* use a word,' Humpty Dumpty said in a rather scornful
> tone, 'it means just what I choose it to mean – neither more nor less.'
> 'The question is,' said Alice, 'whether you *can* make words mean
> different things.'
> 'The question is,' said Humpty Dumpty, 'which is to be master –
> that's all.'

For Humpty Dumpty, the matter is about who is in control, who is in power,
and that's *all* it is. Some of us might like to say 'Who is controlling the narrative', but I have never thought that narrative was what was involved

◆ 'not reading Herakleitos' is not the same as 'not reading Greek' – in some
respects I am feeling as if I have not been all that bright in the effort so far –
I have struggled to achieve, for instance, some insights that are simply everyday understanding among the scholars – Ted Jenner would have told me so
in a flash! nevertheless, I have had to find these things out for myself, & in
such a way that I can be clear about what the issues are, and are going to be,
for me as I engage with whatever 'Herakleitos' I can come to know

◆ on another hand, it's easy to see how combative is a lot of scholarly commentary on the preSocratics in general and Herakleitos in particular – e.g.,
writing on his Fragment 109 (Diels 87), Marcovich seeks to remove himself
and Herakleitos from the prospect of being one of the 'foolish men' or 'stupid
men' who are 'in a flutter at every word' (Freeman), or 'get astounded at every
(new) word' (Marcovich), by describing the foolish one as 'a stupid man
(probably one of οἱ πολλοί)',        *o dear, hoi polloi, c'est moi!*        but when
one looks at the entire field of preSocratic studies, and the incredible detail
some of the philological problematics entail, it's hard not to see the field
itself as a kind of attentiveness that is inevitably 'in a flutter at every word' –
in that sense, Marcovich is a fully paid-up member of the tribe from which
he wishes to distinguish himself – even the briefest acquaintance with
recent studies on the Derveni Papyrus[2] should be enough to demonstrate
the point – that being in a flutter at every word is actually part of the scholarly deal, the intellectual contract that all readers in/on any special subject

matter are implicated in, per se – it is perfectly possible to read the Herakleitean saying as a critique of the very activity of the scholars who wish to be credible witnesses – *c'est moi!*

♦   what can it mean to talk about any writings at all as being preSocratic – *who exactly, before Socrates, thought like him* – for it is hard to think that the thought of Herakleitos, or of Parmenides, was in any way completed or fulfilled or even forwarded in Socrates – and as far as anyone knows, Socrates never 'wrote' anything at all – the 'writer' in this case is Plato, in which case, should not the idiom be 'prePlatonic' rather than 'preSocratic' – if many of the *dialogoi* are presented as records of a past conversation, then are we not dealing here with copyists, scribes, amanuenses, all of whom are then bound into the pattern of *saying what someone else was said to have said,* which is exactly the situation of the transmission of the words of Herakleitos – the danger, it seems to me, of the term 'preSocratic' is that it sometimes carries with it a notion of an early stage in development – that they were precursors to what we are doing now – but the idea that Herakleitos is a sort of primitive precursor to what I'm doing on the other side of the world in the twenty-first century is unacceptable – Heidegger himself issued the appropriate warning, saying Herakleitos is already ahead of us – this very exercise, the writing I am doing at this moment, is a poor attempt to catch up with him – one might as well suggest that we can learn nothing either of the lineaments or the wit of thoughtful feeling, or of the technical skill in writing from reading Sappho, or Simonides, or Archilochos –

the problems associated with the possible abandonment of the term 'preSocratics' have been very nicely articulated by Andrei Lebedev –

> The validity of the term 'Presocratics' has been with good reasons question[ed] by Salomo Luria (С. Я. Лурье) starting from the 1920s (1970: 5 ff.) and by Tony Long in his preface to the *Cambridge Companion to Early Greek Philosophy*. Late Martin West, after reading with approval my (2009) paper 'Getting rid of the «Presocratics»' replied: 'What you say about the Presocratics corresponds to what I have always thought. Forty-six years ago I wrote (CQ 17, 1967, 1 n. 2): "The term 'Presocratics' has so established itself that we should greatly inconvenience ourselves by abandoning it now.

But it has two grave disadvantages: it exaggerates the effect of Socrates; and it lumps together an assortment of people, priests, doctors, vagabond poets, experimental physicists, whose methods and intentions were very various, and implies that they were somehow united in a common search". (letter to Andrei Lebedev from March 2, 2013).[3]

# is it a poem / is it a prose
## is it a writing / is it a saying

Most ancient Greek texts were composed for an audience of *hearers*, not *readers*. As a result, the writing of Greek historically functioned as a recording system of the human voice than a written system independent of speaking.

*Wilfred E Major and Michael Laughy*[1]

♦   Eva Brann's 'prose that could contend with poetry'[2] has all sorts of ripple effects when talking about Herakleitos — part of his text proposes that all things are in contention : 'all things come about by strife' (Loney citing Robinson translating Aristotle citing Herakleitos) — what if Herakleitos was not much impressed either with the poetry or the prose of his day, and Brann's note that he was the 'inventor of the first of philosophical genres, the thought-compacted aphorism' is an important guide to whatever it might have been that Herakleitos 'wrote' — he certainly was familiar with other 'writers' : he mentions many, including Homer, Hesiod, Archilochos, Pythagoras, Xenophanes, Hekataios, poets and prosists alike, & all in order to condemn them — present-day authors tend to see these dismissals as dismissals of their 'theories', but what if there is something in the ways the ancients actually constructed their works ('poetry' and 'prose') that he objected to — in our own time, the Language Poets objected to exactly that, that the normal practices of making a poem and constructing a prose were both to be questioned and bypassed to reflect a need to write in forms very different from them both, precisely because they saw that the social, political, & philosophical assumptions *embedded in those writing practices* were to be rejected — a similar impulse in Herakleitos might be why he went for the 'thought-compacted aphorism' which was neither poetry nor prose — or, what if his own sayings come about by the strife between poetry and prose which engenders a 'third term', the 'thought-compacted aphorism', which can only be heard as an oral- or phonetic-based utterance, & which tends to defy the extended commentaries that can be expressed 'in other words' — the prospect is not without parallels : poet Ron Silliman[3] identified and propounded a new kind of 'sentence' in contemporary writing in which the very values of tradition and power were questioned & replaced with other possibilities : in Robert Bringhurst's essay 'Raven's Wine Cup'[4] he writes of the 6th century CE Chinese poets, saying 'With them, the escape from narrative is complete' : in Ezra Pound's Canto LXXXI he writes, '(to break the pentameter, that was the first heave)' – one such 'heave' was that of Gerard Manley Hopkins and his 'sprung rhythms', or Walt Whitman with his speech rhythms having nothing to do with iambic pentameter at all – but the point is the same : writers have often sought to run counter to prevailing modes of composition in order to realise their own potential, their own works, their own writing – in my own work as a poet, a considerable part of the enterprise thruout the 1970s and 1980s was to resist the then conventional requirements and customs that applied to any writing that sought to get funded, or published, or get a decent shake in the media – it is easy for me to accept that Herakleitos could have been in contention with the compositional conventions of his day – on another hand, it must be acknowledged that not all the sayings of Herakleitos that we have are strictly phonetic-based – many are clearly in the semantic-based writing of the sources – it may well be a trap to suggest that phonetic-based sayings are therefore authentic and semantic-based sayings are not – there's no reason I know of that would propose that every reported saying of Herakleitos is or should be phonetic-based – that some are and some are not can merely reflect either the style of those whose report those sayings, or that Herakleitos himself wrote variously from one base to another – such questions must, it seems to me, remain open –

even so, let me quote here the English poet James Reeves : 'One fact which emerges from the history of poetry is indeed the absolute necessity for *a continual renewal of technical resources* [my italics]; again and again one sees how a situation in which poetry has become sterile, and sensibility atrophied, is changed by a return to a diction and a rhythm more in conformity with common speech. ....There are occasions when a poet can only revitalise his diction by detaching himself from society...'[5] – it seems to me that, among the things that Herakleitos was involved in doing, 'revitalising his diction' was uppermost in a great many of the sayings that have come down to us

♦  was Herakleitos literate – did he write in his own hand – or did he have a slave or scribe to do the manual writing for him – there are plenty of accounts of people, including slaves, dictating letters and other material to a scribe who wrote them down – in discussing the unusual rhythms & syntax of the sayings of Herakleitos, Eric Havelock[6] has proposed that 'statements of this type were framed not to be read but to be heard and memorized' – he goes on : 'The fact that other kinds of rhythm are substituted [i.e. for those of traditional poetries] should not obscure the essential point, that these rhythms are less regular, or more precisely that the pattern within any given statement is unique. One line in this style is not a variant of any other. The hexameter flow has been broken. …Each statement therefore ends up as self-contained and autonomous. You cannot add to it or subtract from it'. Havelock's case is that the sayings of Herakleitos are oral ones, to be verbally delivered to an *audience* rather than written or published for a *readership* – looking at the internal evidence of the texts, there is only one fragment that appears to refer to 'writing' (ie γράφω, which refers to graving on stone as much as to writing on papyrus – '(1) To scratch, cut into, incise… (2) To form by scratching or incision' – from Cunliffe : γραφέων ὁδὸς εὐθεῖα καὶ σκολιή – 'the way of writing is straight and crooked' (trans Robinson) – where one only has to look at the Greek alphabet, all in capitals at that time, to *see* the straightness and crookedness of the actual marks on the page or stone or pot – R P Austin[7] pointed out (1938) that in the Attic alphabet in the late 6th century BCE (when Herakleitos was born) were 'fifteen letters which contained as part of their forms the vertical stroke or its near neighbour the slightly oblique stroke' – apart from Θ, Ο, Φ, Ω, and the sign for *koppa*, all the inscriptional letters of ancient Greek were angular, straight and crooked, with just five letters with rounded forms out of approximately 24 letters – for the rest, there are 13 fragments, about 10% of the total number, which explicitly refer to *hearing* and *listening* as a means whereby the λόγος can be perceived or understood – (in Robinson they are fragments 1, 19, 34, 50, 55, 74, 79, 87, 92, 101a, 104, 107, 108) – the words of Herakleitos may be thus understood to be *sayings* rather than *writings* –

again, Havelock is sharp and down to earth in these matters : 'Out of a total of some hundred and thirty sayings, no less than forty-four, or thirty-four per cent, are preoccupied with the necessity to find a new and better language, or a new and more correct mode of experience, or are obsessed with the rejection of current methods of communication and current experience.'[8] – he then lists all 44 of them, and says 11 more 'could be added…

if viewed in the same light'[9] – Robinson's Fragment 101a is hoisted into view : 'Eyes are more accurate witnesses than ears', and Havelock comments: '…if the advantage of eyes over ears then be taken to refer to reading (his sayings) as against listening to a recital' – what I hear here is that the communication of 'text' is still an oral matter within the culture, while Herakleitos is acutely aware of the capacity to think more clearly if one can have the words in front of you – listening is what everyone has available to them, but reading allows a more accurate attention to those texts – in our own time, Charles Olson put it beautifully : 'The advantage of literacy is that words can be on the page'[10]

♦   yet, to have the same number of 'readers' as one could have in an 'audience', there would have to be more manuscript copies of the text available than there could be in any audience or succession of audiences – well, that's possible with printing, but not with manuscript copying – it raises the question : did *anybody* read (in the sense in which we today mean 'read') Herakleitos in his own time – I have found it impossible to find out what 'reading' might have meant in terms of how it was done in the period – of the later 12th century, Paul Saenger points out that Hugh of St Victor (d. 1141, Paris) 'theorized about the new modes of reading canonically separated script. In the *Didascalicon,* he explicitly set forth three modes of reading: reading to another, listening to another read, and reading to oneself by gazing (*inspicere*), that is, silent, private reading'[11] –

it seems reasonable to guess that these options were also available to a 5th century BCE reader, and that 'reading' as a means of following what one is hearing, 'listening to another read', may have been a contemporary practice – Saenger's general proposition is that silent reading is a function of the historical separation of words in the line – if he's right and, taking it that ancient Greek text did not separate words in the line, then reading as reading aloud (to oneself, or listening to another) may well have been a norm at that time –

it is certainly possible to think of a few readers without expanding that into our own cultural capacity for 'readership' – for we have no evidence at all of copies made for others to read, tho it seems reasonable to say that someone must have recorded what they heard, or we would have no fragmentary 'writings' to go on – is it even possible to ask : Who first recorded the sayings, who copied them, and whose copies at how many degrees of separation got into the hands of Socrates or, a few hundred years later, Clement of Alexandria or Hippolytus of Rome – and yet the modern literature abounds with declarative statements about what Herakleitos said, or should have said, or meant, or *really* meant –

AND – who would have had access to that much papyrus – making papyrus sheets was time-consuming and expensive – our modern habit of 'printout' and 'handout' is not feasible until the 20th century and the appearance of the first commercially available photocopier, the Xerox 914, in 1958 – papyrus was never made in Greece – not even scriptoria, as places set apart for writing or copying, are attested in fifth century BCE Greece, they seem to have been a specifically christian mechanism and began around the 5th century CE, some 900 or so years after Herakleitos –

it is not so much that Herakleitos did not 'write' anything, there is simply no evidence that he did – but of course that can be changed by new discoveries – it is time for Keats's Negative Capability to kick in, that we can be 'capable of being in uncertainties, Mysteries, doubts, without any irritable reaching after fact & reason' – tho it is interesting that the search for 'fact & reason' is 'irritable', that Keats regarded such efforts as emotional – that the very search for 'fact & reason' might not spring from fact or reason – (letter to George and Tom Keats, December 1817)[12]

♦   to return for a moment to the questions of hearing and listening, here's Robinson in his commentary to Fragment 50 : 'one does not 'listen to' patterns, or structures…one listens to persons, and the things they say' (p114) – but Robinson himself notes puns thruout Herakleitos – to which Havelock adds (tho long before Robinson) 'repetition, assonance, antithesis, and symmetry' – all of which affect deeply the sheer *sound* that the texts of Herakleitos make when one reads the words aloud – when we say that much is lost in translation, we do not usually detail exactly what is lost outside of questions of nuances of meaning, but it is clear to me that part of what is lost is the actual noise the saying makes in the original language, which cannot be replicated in the language of translation – and for this poet, it is part of the sensible universe that the sound that words make is part of their total communicative function and effect – and if that were not so, why on earth would

poetry have developed in the first place, where part of the precise purpose
and pleasure of the poem is (back to Robinson again) to 'listen to patterns or
structures' – the English sonnet e.g. is an almost fixed structure, where both
poet and reader/listener know where the lines end & the whole poem itself
ends – not only do we hear the rhymed line-endings, we anticipate them,
and we sense/know when the last or even second-last line of the poem is
arrived at – classicists have told me that the same applies to how they read/
hear archaic hexameters : give them a part, and they hear the rhythm of the
rest of the line – New Zealand poet C K Stead once wrote about how readily
he was able to think in the sonnet's 14-line chunks once he embarked upon
a long sequence of them – we do listen to and for patterns, and Havelock
makes a case I think for these patterns being critical for Herakleitos when he
abandons the usual compositional structures of the poetry or the writing of
his day – Robinson's Fragment 10 is a fine example of what I'm getting at –

συλλάψιες: ὅλα καὶ οὐκ ὅλα, συμφερόμενον διαφερόμενον, συνᾷδον διᾷδον
ἐκ πάντων ἕν καὶ ἐξ ἑνὸς πάντα

here are the sounds : hola/hola, omenon/omenon, adon/adon, panton/
panta, en/ex, hen/henos – the vowels repeating with slight variants, but
always each sound supports and enhances other sounds in the lines – if all
one wanted was meaning, then there'd be no need for such verbal pyrotech-
nics – but if what one wanted was to remember the words, then *the words
themselves become their own aides-memoire* in the sayings of Herakleitos –
& as with the 6th century Chinese poets, 'the escape from narrative is complete', as in my 'translation' here (adapted from Robinson's 'Things grasped
together: things whole, things not whole; <something> consonant, <some-
thing> dissonant. Out of all things <comes?> one thing, and out of one
thing all things.' –

taken together                         whole & not whole
             together              separate
       consonant        dissonant
              from all              one
       from one          all

I have left Robinson's commas in the Greek (they are his, there are none in ancient Greek), but not used them in the English – it's important to remember that ancient Greek had no punctuation as we understand it – *printed* Greek has, and has had it since Henri Estienne's first printed edition in 1573 – Bywater, Diels, and everybody since uses commas, full stops, etc in their editions – except Davenport, who employs line breaks to separate meaning units (if I may call them that) – but in spite of the fact that such punctuational usage has become conventional in printed texts, it still constitutes a mode of *editorialising* in texts that never had punctuation in their originals – we know that in English, a comma can alter meaning ('eats, shoots, and leaves', instead of 'eats shoots and leaves'), and in Greek it is the same : the insertion of a comma or full stop is an *editorial* decision taken by someone who is not the author, after the *compositional* decision has been made – in other words, the commas do not belong to Herakleitos, but to the editors, or perhaps more accurately, to the entire ethos of classical studies, literary and philosophic, since the birth of printing – nevertheless, what I want to say here is that Robinson's Fragment 10 is one of the primary descriptions of ὁ λόγος

this section is a selection of quotations from various writers whose studies
have helped me to have a wider view of the epigraphic material that provides
contexts for the original writings – they treat of matters more specifically
verifiable than any Herakleitean text, & I am interested in how these infor-
mations have affected my approach to what Herakleitos may have said –
they are not in chronological order, but I have sequenced them in a kind of
order, for which I really have no name – not ordered, not random – if anything it is that I have come upon these things & I wish to see them in a way
that contributes to my own understanding, still in process, of the appearance
of a developed Greek alphabet and its occurrence on various surfaces, stone,
pottery, tablets, papyrus, circa 700 BCE to 401 BCE – the dates of Herakleitos
are not exactly known but are generally accepted as about 535 to 475 BCE,
crossing from the 6thC into the 5th –

from Stephen V Tracy[1] –

in his 1990 publication he notes : 'when I began my work twenty years ago,
it was generally denied that one could identify the work of individual cutters'
(pxv) – Later, on p 3, he writes :

'Learning' the hand was and is the real challenge in achieving any
meaningful study of ancient cutters. One does it by repeated study
of the lettering on the fragment selected, drawing every letter, sometimes over and over again, and noting every variation. It is a slow,
painstaking process requiring weeks and sometimes months; the
time varies from cutter to cutter. After considerable study it helps to

verbalize, that is to put in writing, the peculiarities of the cutter's style. By slow degrees one gains a familiarity with the lettering until at some point (I do not know exactly how) one comes to 'know' the hand just the way one knows the writing of a close acquaintance.

Tracy also elaborates a further complication in the identification of hands: 'The only valid criterion for calling one cutter an apprentice of another can be, it appears to me, a significant similarity, not merely in the shapes of the letters but in the personal, often trivial, idiosyncrasies that set one cutter off from another.' In a foot-note he adds, '…it is possible for an apprentice to study under a cutter extensively and still cultivate a style of lettering very different from that of the master. In that case, we have no way of recognizing the relationship between them.' (p230) –

it should be added that *recognizing* a hand and *identifying* that hand are not the same, as anyone could attest who has studied the handwritten letters, diaries, notebooks etc of other authors –

Tracy's work covers the years 229–86 BCE, somewhat later than my general focus here, and is confined to texts cut in stone, but it is clear that the identification of a 'hand' requires a lot of study and a lot of examples to compare with each other – Tracy's list of fragments 'not assigned' (i.e. to an identifiable cutter) is not only considerable, but are all either unique examples (only one fragment by this cutter) or not enough examples to establish an identifiable style –

recently, I contacted Professor Tracy & asked whether he thought the early stone-cutters were literate or not – his initial view was that they were, and yet he conceded that 'In principle, I can think of no impediment to an illiterate cutter simply copying letter shapes. As he copied his eye could still skip ahead etc. resulting in the haplography [a copying error where the eye skips from a word in one line to the same word in a later line, omitting the words in between]. With short texts, this strikes me as entirely possible. It is harder for me to imagine in the case of long decrees inscribed with thousands of letters 0,008 m. [8mm] in height. Such texts are the norm in Athens from IV BC until the Roman period'.[2] The prospect is, then, that very early inscriptions could have been made by cutters who could not read at a time when literacy was not so widespread – and while that seems sound enough, we don't have enough information to make declarative statements on it – Tracy also told me that 'copy' for ancient inscriptions is simply not known, so we don't know what kind of documents served as the 'originals' for the stone-cutter's work, nor how they were used in the cutting process – L H Jeffrey's examples of stone inscriptions from the 7th to 5th centuries BCE show relatively crude cutting or incision compared to the more regular & standardised letterforms of the 4th and 3rd centuries BCE, and I wonder if the progress toward a more sophisticated lettering was accompanied by a greater degree of literacy across the general population – in any case, writing in the time of Herakleitos pre-ceeded the regularisation of the alphabet in Athens by at least 70 years

from R P Austin[3] –

p 1 – the stoichedon style of lettering is a style '…in which the letters are in alinement vertically as well as horizontally, and are placed at equal intervals along their respective alinements'

p 6 – '…the stoichedon style is known to have come into being not later than the second half of the sixth century BC'

that is to say, during the time in which Herakleitos was born – tho I must say that stoichedon is almost completely confined to Attica, & Herakleitos was almost completely confined to Ephesos on the eastern Aegean coast – the point I am heading towards here is not that Herakleitos knew anything about stoichedon, but that stoichedon developed at the same time as that boustrophedon writing was giving way to prograde writing – in other words, during the life of Herakleitos, the manner of writing, the orientation of the letters, were still in a state of flux, which is easily seen from stone inscriptions, writing on pottery and tablets of the time –

p 17 – 'In the middle of the sixth century Greek writing…was in a transitional state. The great change from retrograde to left-to-right…had so far progressed…that boustrophedon, may be considered typical of the period… Engravers seem to have written retrograde, boustrophedon, or left-to-right as suited their fancy…Boustrophedon…was dying out in the second half of the sixth century, at which time…the stoichedon style came into being.'

stoichedon was incised or cut into stone on a grid (Austin called it a 'che-quer') – a grid of vertical and horizontal lines which were often drawn onto the stone, the letters cut in the centre of the squares of the grid, then the grid erased to let the letters stand on their own – whatever else such a process achieves, it requires a considerable uniformity of letterform and letter size for it to work –

and yet, alongside this growing uniformity, many inscribed letters on stone, pottery, and tablets, exhibited a decided lack of uniformity across the various mediums and thruout the 6th and 5th centuries – even so, L H Jeffrey makes the point that boustrophedon was relatively easy to read because 'twelve of the twenty-six shapes were symmetrical…six required very little change… and only eight looked markedly different in reverse…'

alongside this, inscriptions incised or painted on what we know as Little-master cups contain inconsistent letterforms on the same cup – examples are known with different forms of sigma and lambda, for instance, and with both prograde & retrograde forms – these cups are all wide-mouthed, shal-low goblets, the tallest of which seem to be about 14cm and others as small as 8 or 9cm – the band around the bowl where lettering often occurs is there-fore a narrow one, in which the artist is required by the space available to write in letterforms of low and uniform height – coincidentally, these cups only, like stoichedon, appear in Attica, and only, like stoichedon, appear in the latter half of the 6th century – I do not know if these phenomena influ-enced each other – but it is true that the writing of text in the late 6th century was undergoing a somewhat sputtering shift from boustrophedon to pro-grade writing, & another shift from irregular-sized to uniform-sized letter-forms, and that this shift affected all types of inscription of the period and of the place – [see p 37, Pieter Heesen, *Athenian Little-master Cups* : https:// pure.uva.nl/ws/files/755640/70414_text.pdf] – &, for my purposes, this shift also coincides with the dates of the life of Herakleitos – so, if he did write by hand, what might his writing have looked like

◆ as an aside [there are only asides! one of the chief myths about Herak-leitos is that he 'wrote a book' – and here I must invoke a bit of data from my other deep interest : the book itself –

whatever it is that anyone writes, it is not, and never can be, that they are writing a *book* – people write poems, letters, short stories, memoirs, biographies, histories, articles, essays, theses, dissertations, novels, diaries, philosophies, plays etc, but they never write a book – books are manufactured by those who publish those writings in book form – of course, the claim is made everywhere : 'well, you know, when I was writing my book' – but a text does not become a book until it is published – and publishing means what it says, a writing is made available to the public by a multiplication of *copies*, not a multiplication of *originals* – a private text is made into a public service we call 'book' – if such recording/copying was done on a roll or scroll or tablet in ancient Greece, there is no evidence for the existence of any such roll or scroll or tablet of work by Herakleitos – these are very specific objects for which we have long had words, namely, 'roll', 'scroll', and 'tablet' – why call them 'books' as well, is beyond me, in a discussion where, in all other matters, careful distinctions are forever being made, questioned, reconfigured, & made again – historically, the book came about over half a millennium later than Herakleitos in the 1st and 2nd centuries CE – Coptic monks in northern Africa and southern Italy invented the codex as a much less cumbersome instrument for gathering & storing texts & getting access to them for reading than the roll or scroll or tablet could ever be, including the ability to write on both sides of the papyrus – it is this codex (which refers simply to the *structure* of the object : a bunch of papers or other materials joined down one side with a protective cover) plus its textual and visual matter *taken together* that has become known to us as a 'book' – whatever Herakleitos might have written, he did not write a book

L H (Ann) Jeffrey [4] –

meanwhile, back at the alphabet – in the last hundred years or so, an extraordinary amount of research material has been published on the origin and spread of the Greek alphabet – yet there seems to be a consensus that the alphabet was formed largely from the consonantal, retrograde north Semitic alphabet, tho there is some question around just how the vowels came to be established – Rhys Carpenter is adamant that the vowels were adopted from the Cypriot Syllabary (an alphabet comprised of syllables rather than letters, and where each syllable was made up of a consonant followed by a vowel), but others derive the vowels from adaptations of other Semitic letters for the purpose – another consensus is that the alphabet was formed in or by the eighth century BCE, & the first known examples of its use are found incised on stone, marble, and pottery during the 700s –

Ann Jeffrey published in 1960 a brilliant survey of the local scripts of Archaic Greece from the 8th to the 5th centuries, and her work was updated after her death in 1986 for a new edition in 1990 by A W Johnston – Jeffrey showed examples, not only of variations in Greek letterforms from one locality to another, but also an often wide range of variant letterforms *within* a particular locality over time – when we take into account the official standardisation of the alphabet in Athens in 403 BCE, it is easy to see that geographical and temporal variations in the forms of letters was the norm from the 8th to the end of the 5th century – 400 years of alphabetic instability –

looking just at Central Greece, Jeffrey found 7 different forms of alpha, 3 of beta, 4 of eta, 3 of theta, 3 each of kappa, lambda, and mu, 4 of nu, 5 of rho, 3 of sigma, 6 of upsilon, and 3 of phi – with one version only of zeta, iota, xi, omicron, pi, koppa, tau, and psi – besides 'Central Greece', Jeffrey's localities are listed as The Peloponnese, North-Western Greece, The Western Colonies, The Aegean Islands, The Eastern Greeks, and The Northern Colonial Area – with a total of approximately 100 specifically named places, all with variant letterforms between one place and another and within their areas – in the category of The Eastern Greeks, which includes Ephesos where Herakleitos apparently lived all his life, Jeffrey's examples are : a silver plaque recording contributions of gold and silver c550? – fragments of dedicatory inscriptions on lower column-drums c550?, and two blocks of stone dealing with augury and oath-taking c500–475? – all within the lifetime of Herakleitos, with nothing in the way of literature or philosophical speculation – to be sure, a few pieces in Ephesos is not a lot of evidence, but taken across Jeffrey's entire Greek local coverage, the number of non-papyrus written artifacts runs into the hundreds – that is, not much evidence for a local written literary or philosophical context in which Herakleitos might have lived and worked – it is hard to imagine what sort of animals 'the writings of Herakleitos' might have looked like

*A N Whitehead* : our single datum is the whole world, including ourselves
*Charles Olson* : the 'subject' is the total on-going reality as of now
*Jan Zwicky* : the lyric is an attempt to express the Whole in a single gesture
*Samuel Beckett* : an art that comes complete with missing parts
*Ludwig Wittgenstein* : how small a thought it takes to fill a whole life

*A Cautionary Tale – the appearance of further discoveries*

Robinson Frs 3 (Aëtius) and 94 (Plutarch), both in 1st and 2nd centuries CE –

Fr. 3 : (περὶ μεγέθους ἡλίου) εὖρος ποδὸς ἀνθρωπείου – [The sun's] breadth is
<that> of a human foot.

Fr. 94 : ἥλιος γὰρ οὐχ ὑπερβήσεται μέτρα · εἰ δὲ μή, Ἐρινύες μιν Δίκης ἐπίκουροι
ἐξευρήσουσιν. – The sun <god> will not overstep <his> measures. Otherwise <the> avenging Furies, ministers of Justice, will find him out.

since the mid-19th century, these two sayings have been published & treated
as being entirely separate within the accepted oeuvre – Robert Creeley once
said in an interview, 'Context is everything', so in that sense, a quotation
delivered without its context may lack 'everything' of importance, & all we
have is a few words, which we have tended to take at face value – the first
fragment has 'the sun's width' as its subject, the second deals with the consequences of the sun overstepping its measures – in the editions I have, none
links the two sayings into a single statement – when it comes to 'context', the
first and immediate context of any saying or written text is : all other written
words around them –

the Derveni papyrus brings both sayings together as the immediate context
in which each has its life and meaning – of course, scholars have differed in
how Derveni is to be translated, in part because the condition of the papyrus
is seriously compromised by having been a component in a funeral pyre –
it's a wonder there's anything of it left at all – & what is left is a lot of charred

material which has had to be x-rayed and microscopically examined to make out the remaining text – so speculation about what's missing has been very active and has a lot of scope – one scholar, Walter Burkert[1], advises : 'One new letter added or clearly read may change the whole. It is good to remember how for Antiphon F 44 DK, a comparatively simple, unpretentious text, one little piece of papyrus added later (1984: *POxy.* 52 nr. 3647) brought half a dozen of corrections.' – and he offers a tentative translation of the Herakleitian text being in turn a quotation by the Derveni author : 'Helios, according to his own nature, a human foot's breadth boundaries not surpassing; for if he will step outside his own boundaries, Erinyes will find him out' –

the character of the Derveni quotation has thus changed from its usual presentation as two separate and unrelated sayings into a single statement about the boundaries of each thing's nature – it is also a very small text that has contradicted over 2000 years of time-honoured tradition in which a large number of words have been expended on incorrect or misleading readings – it may be an exceptional case, we don't know, but it is definitely some sort of index of the extraordinary fragility, not only of the physical evidence, but also of the readings that can be made of it

♦  gathering groups of sayings under 'subject headings' risks treating the saying as a function of the subject heading rather than as a text having its own weight, heft, and significance independently of one's own professional status, activity, protocols, & concerns – presumably, the tendency of some scholars to gather various sayings under a sort of subject-heading is the hope that a number of disparate statements might, if they were *all* able to be gathered, the lost ones freshly discovered, make up a continuous story or narrative, or elaborate an integrated philosophical 'system' which the isolated sayings do not permit, a bit like a jigsaw puzzle with many of the pieces unable or yet to be found – just as Herakleitos did not write a 'book', the question whether he wrote a unified philosophical narrative must also remain open, with the prospect that, in the light of further evidence, we might discover he *never* wrote a unified 'work', as many of us tend to understand that term, at all – or alternatively, our conceptions of what is properly able to be called a 'work' might have to undergo considerable rethinking – in any case, those subject headings are inventions of the editor, the scholar, the commentator, even the publisher, but not of the author – a few examples will do –

a) Aulus Gellius[2] (Rome, 2nd century BCE) born in Rome, spent some time as a young man in Athens, and while there began a sort of diary, 'a collection of interesting notes on grammar, public and private antiquities, history and biography, philosophy (including natural philosophy), points of law, text criticism, literary criticism, & various other topics', says his early 20th century editor & translator, racking up a considerable line of 20th century subject headings in order to show the variety of Aulus's 'interests'. Aulus himself had a different sense of what he was doing. Of his method he writes, '... in the arrangement of my material I have adopted the same haphazard order that I had previously followed in collecting it...I used to jot down whatever took my fancy, of any and every kind, without any definite plan or order' – we all know of this kind of process, many of our own notebooks and diaries will follow just such a procedure, and no one to my knowledge allows Herakleitos to have had any like process in mind –

b) Leonardo da Vinci[3] (Italy, 1452 – 1519) wrote some of the most absorbing and detailed notebooks of anyone who has lived – in facsimiles of them it's easy to see that his attention was taken up by one thing after another on a single page in a display of diverse attentiveness barely matched elsewhere – but translators of his notebooks have been unable to let Leonardo 'have his head' in these matters, Edward McCurdy (1938) writing in the Preface to his translation : 'Some classification of the material, however, has been found to be necessary on account of the extraordinary diversity of the subjects treated of in the same manuscript, in the majority of cases' – in contrast, I would have thought that that 'extraordinary diversity' was exactly what ought to be maintained to be truthful to the actual operation of Leonardo's process – instead we get our modern specialisations of 'Philosophy', 'Anatomy', 'Astronomy', 'Optics', 'Physical Geography', etc, as if our linguistic categories can be inferred backwards & visited on unsuspecting predecessors as if it does not matter or makes no difference to our understanding –

c) Joseph Joubert[4] (France, 1754 – 1824) wrote a series of notebooks, writes his translator Paul Auster, 'as a way to prepare himself for a larger, more systematic work, a great book of philosophy that he dreamed he had it in him to write. As the years passed, however, and the great project continued to elude him, he slowly came to realize that the notebooks were an end in themselves. By 1804, he was able to admit that 'These thoughts form not only the foun-

dation of my work, but of my life'. ' – the order of the notebook entries
therefore was the sequence of the years in which they were written, with
no attempt to incorporate them into any other story or narrative which an
editor may have felt was 'necessary' – Paul Auster's wish to allow Joubert to
speak as he wrote is a courtesy, it seems to me, of a high order – what would
happen to Herakleitos if the same courtesy was extended to him – that no
attempt was made to shape or reshape his work until an ur-text or an arche-
typal text or a great many more fragments were discovered –

d)   Louis Zukofsky[5] (USA, 1904 – 1978) once suggested that a poet writes
one poem all their life – whatever the variety of works, their techniques,
their multiplicities, they add up to a single attention, a single enterprise,
a single work – he writes (in 'A-12') –

> Each writer writes
> one long work whose beat he cannot
> entirely be aware of. Recurrences
> follow him, crib and drink from a
> well that's his cadence – after
> he's gone.

'his cadence', as we all have one, whatever its nature, whatever flows from it,
whatever diversity issues from it – it is the same well or reservoir that the
poet or philosopher drinks from all their life – even in such simple things,
like 'knowing' what someone will think of this or that matter – it applies to
Herakleitos as much as to Zukofsky – the prospect most decidedly exists
that Herakleitos wrote one work all his life – and that the apparently fragmentary nature of the bits we are permitted to glimpse thru the haze of the
words of others is all that we were ever going to get, *even if we had every one
of them before us* –

it is also clear to me that there is something odd about how Herakleitos's
work gets to be understood : it seems to be assumed that he wrote only once,
at one point of his life, wrote a short, single 'complete' work, and then nothing for the whole of the rest of his life – there is no sense at all that I can see
that he may have written *this* when he was 25 years old, & *that* when he was
47 – no inkling that his works may have been somewhat semantic-based

early in his life until he developed a terse and compacted mode of writing in a phonetic-based manner later in his life – because it is certainly the case that, taking it that (for the moment) he was quoted accurately by the early recorders, some of his sentences are semantic in mode and some are phonetic in mode – I have seen no such observation, except by Eric A Havelock, in modern writings – but could there be a young Herakleitos and an older Herakleitos *in the sayings* –

e) Alan Loney – here is a page from a recent poem[6] –

utterance = saying something    stutterance = not saying something

poetry = utterance&stutterance          revealing  /  concealing
                                                           repeating  /  repealing

what is it each
of us is not
permitted
to say

*when told I am as mad as a snake*
*I was relieved and have breathed*
*more freely since that day*

'next to nothing' – less
a quantity than a location
there, alongside you

when I sent some similar material to a publisher about ten years ago, they replied that they'd like to see these 'fragments' worked up into fully developed poems – their assumption was that here were notes made in haste that needed more time to grow into more fully integrated and elaborated wholes – my reply that these notes *were* the fully developed wholes, that these little clusters of words were the result of a careful working and reworking of the material that came to me was too much for that otherwise perfectly sensible and experienced publisher to accept

♦   we say, among the things we say, that what we 'have' (whatever 'have' can mean in such a context) only *quotations* from our wily philosopher –

but what do or can we mean by 'quotation' – we tend to use it as if the meaning is obvious and straightforward to everybody, and that other people will understand without us having to take any time to clarify our usage – so, what acts do we perform by using the term – there are four aspects of this 'problem' I wish to highlight –

1) *our carefully crafted texts vs publisher's house-style* – many of us have had this problem – that aspects of our text are changed because the publisher wants a uniformity of practice in their publications when it comes to such things as how dates are to be written, how our references are to be constructed, how commas will be treated in lists within a sentence, among other things – there is probably not an academic or university press anywhere who would publish my prose as I have written it here – without question marks, or sentence markers like capital letters and full stops – and indeed I declined an offer from a university press a few years ago because they would not accept the way I made paragraphs (I spent about 20 years developing it) and wanted to convert my writing to standard expository prose – we are so used to these processes that we have often lost the sense of their being intrusions into our texts, into our ways of thinking, into our ways of expressing ourselves – [and brackets and parentheses that do not close are unforgiveable for some editors and publishers –

we also know that, from the beginning of publishing in the West, publishers & typesetters have altered their texts to conform with other notions of how a text is supposed to be 'presented' – the printing and proofing practices of 18th century printers have already been shown by Randall McLeod to have produced multiple versions of some passages within the same edition of the book – when I worked as a proofreader in newspapers in the 1960s, many Linotype operators used H W Fowler's *A Dictionary of Modern English Usage* to settle questions raised by the copy they were busy setting in type – these were often questions of *general usage* – questions of *specific usage*, were also settled by reference to a publisher's own style book, which selected specific solutions to a range of options which general usage had available to it –

it is one thing to say, well, these changes are merely matters of 'punctuation' or 'presentation', not matters of substance – but one of the great insights of 20th century textual criticism is the dispersal of the difference between what used to be called 'substantives' (content) & 'accidentals' ('punctuation' and 'presentation') so that *all* differences between one version of a text and another are deemed to have implications for content and how it is received – house styles may well offer themselves as harmless matters of presentation, but again and again they have been shown to be false hopes when the work is read closely – the parallels between publishers' house styles and scholarly protocols and conventions (*Heraclitus* may be taken as incorrect in terms of Greek usage, but it is conventional in terms of Roman/English usage) are too numerous to be casual about –

2) an editorial problem we do not have often these days was very common during the 20th century when it came to Concrete Poetry – in large measure, Concrete Poetry was made possible by the typewriter, whose letters occupied the same lateral space as each other, rather like the Greek *stoichedon* texts of the 5th century BCE – occasionally an editor accepted a concrete poem for publication, but their house style required everything in the publication to be in a specific typeface, especially one that was not a typewriter font – in the often tortuous attempt to render the original in a non-typewriter type, the visual shape and communicative effect of the original poem was lost – if one was wanting to 'quote' from a concrete poem, one would have to quote the *visual* aspect of the poem as well as the *auditory* one – and the best way to do that would be to reproduce that part in facsimile rather than try putting the 'text' in a different typeface because one values one's house style so highly – there are texts in which the typeface used is central to their meaning – there are texts for which the typeface used is part and parcel of the work to be conveyed – the idea that it does not matter what typeface is employed does work for some texts but is disastrous for others – we can be fairly precise about those that work and those that don't in this regard –

3) *house style/scholarly apparatus* – the iMouseion Project is a great place to see this issue in full flower – the achievements of K Tsantsanglou and G M Parássoglou are exemplary in the presentation of this text, which comes with an extraordinary range of difficulties to be dealt with – I especially like being able to click on the 'blank' part of the page and see all the brackets and what is inside them disappear – initially I thought, here are just the letters that are clearly identifiable, yet the critical apparatus was still visible – *part of that apparatus is the very typeface used to present the original text* – it is a

modern cursive type, nearly all in lower case, with word-spaces – where the
original Derveni letters are in capitals and in *scriptio continua*, i.e. without
word-spaces – my hope was that I would be able to get a glimpse of what one
might see if one could actually see the papyrus – what we do see instead, is
the *scholarly house style* in which all modern classical scholarship is, apparently, expected to be presented[7] –

4) Gregory Nagy of the Center for Hellenic Studies at Harvard has a fine
essay devoted to an aspect of quotation in *Recapturing a Homeric Legacy :
Images and Insights From the Venetus A Manuscript of the Iliad*[8] – Venetus
A was copied in the 10thC from a document that no longer exists – in the
document we have, there is more writing of scholia or marginalia than there
is of the *Iliad* text itself – the top and bottom margins, the outer and inner
margins, are often full of writing, along with other scattered notations
between the lines of the Iliadic text – a word-count comparing the text of
the *Iliad* and the texts of the annotations might well be interesting – add to
that word-count those of all other scholarly writings about the documents
and the ratio of *Iliad* to its commentary would see that of the *Iliad* in a
never-ending shrinkage –

in Venetus A, Nagy identifies four categories of letters – 1) that of the text
of the *Iliad*; 2) those of the scholia; 3) the lemmata in which one assumes
the scholiasts try to replicate the letter of the original document; 4) the presumed original letter from which the *Iliad* was copied –

Nagy's aim in his essay is to identify why the quotations in the scholia are
not in the letterforms of the scholia themselves, nor do they copy the handwriting of the Iliadic text – he makes a brilliant and convincing case for see-
ing the lemmata as guides for how to read the Homeric text aloud –

my own purpose is different, and is about *quotation* and how we can understand this term – it's important to grasp that the Venetus A lemmata are not
the result of a 'cut & paste' process – they are separate letters, freshly drawn,
copying a text we cannot see – and that means we have no way of checking
whether the 'copy' is always and strictly accurate to its original – when we
cut and paste something on the computer from one file to another, the material we cut is often in a typeface and size that is not that of the type and size

of the type in our own document – at that point, we usually highlight what has been pasted, then reconfigure it into the type and size of our document – we do not usually retain the actual visual cut within our own prose – *but if we did*, if we always left the type and size of the cut material to remain in its difference beside our own type and size, then we can say correctly that we have transferred 'these letters' into 'my letters', a lemma in a strict and literal sense of something taken/received from one place and transferred into another without any loss of the integrity of the original text in the transfer – here we are back at the problem of reproducing or quoting from a concrete poem and its transfer from one typeface to another – even an apparently accurate quotation has the capacity to elicit different meanings & responses from different readers according to their individual circumstances and inter-ests – the very notion of 'quotation' is as slippery a process as is the material one may wish to quote – it's not that no translation or transcription can be trusted, but rather that no translation or transcription should be trusted

*Lew Welch* : Guard the mysteries!
Constantly reveal them!

or, perhaps, simply, the back luck section – guilty of what he complains
about – Heidegger's 'interpretation is hazardous' – in any case, a return to
the anonymous seminar participant who said 'More is said in the interpretation of the fragments than stands in them' – what *can* I say – what does the
evidence *allow* me to say – what can I say that does not violate the very text
I wish to understand – this is where depending on the deep end is all a diver
can do –

one cannot interpret The Dark One – one must become The Dark One to
become the One – there's nothing more mysterious or 'mystical' about it
than a tree is mysterious, a stone is mysterious, a river is mysterious, or a
cloud, a person, a book, a bird, or a knife is mysterious – if we have lost a
sense of The Mystery, it is only the loss of the mystery of the local geography (Charles Olson), the familiar (Herakleitos), the close at hand (William
Carlos Williams), the totality of what we know (Anne Carson), the world of
things (George Oppen), our life of thoughtful feeling (Sappho, Hilda Doolit-
tle, John Keats), the scope of everyday language (Gilbert Ryle, Heidegger,
Gertrude Stein, Wittgenstein) –

Charles Olson :    we are born many
and the single is not
easily known

St Matthew :    if your eye is single
your whole body
will be full of light

♦ περὶ φύσεως (peri phuseose) – these two tiny words, commonly translated as *On Nature* or *Concerning Nature*, and taken to be the title of Herakleitos's work, seem to have achieved the status of a roadblock in Herakleitean studies – so many essays are predicated on the idea that one enterprise of scholarship should be directed to the 'reconstruction' of this imagined 'work', along with the notion that the work is divided into three sections (via Diogenes Laertius) : On the Universe; On Polis; On Gods – & repeated by Andrei Lebedev[1], who further divides the 'work' into smaller sections : 'I. Logos (metaphysics and theory of knowledge); II. Cosmos (philosophy of nature); III. Man. Soul. Life and Death (Physical anthropology and psychology); IV. Ethos. Good and Evil. Arete. (Moral philosophy); V/1. Polis: The world of crafts and arts (texnai) (social anthropology); V/2. State and Laws (political philosophy); VI. Peri Theon (Popular religion and philosophical theology)' – all of these categories are invented by the scholar, all of them are based on a technical language that did not start to be formed until the 18th and 19th centuries – taken together, these 'categories' have the distinct flavour and shape of a completed scholarly book's *Contents Page*, as if the author is making preparations for fitting the as-yet-unfound fragments into the jigsaw puzzle of the yet-to-be-completed work – is already shaping the book to come – Lebedev is not quite alone in this endeavour: Marcovich, and Kirk and Raven, have versions of it –

what if one rejected the very notion that there was *one work* in the first place, or, if one rejected the notion that a 'work' was always and necessarily some kind of unified and consistent whole towards which scholarship can direct its activities – when it comes to it, it may well be that a legitimate object of scholarship is the *reduction in the number of individual sayings so that their individuality is retained,* rather than reducing the number of sayings by joining them up into a single narrative – on the division of a presumed single work into sections with separate subtitles, the efforts of Diogenes Laertius may be no more useful or correct than those of Lebedev – the term περὶ φύσεως is attached to various writings by others (Kathleen Freeman cites Empedocles, Gorgias, Prodicus, besides Heracleitus) suggesting that the term is a general category of writings of a scientific or philosophical sort, rather than a specific title for a single work of a single author – in that sense, περὶ φύσεως might be all that any of us can ever talk about

♦ a major feature of most of the English editions of Herakleitos I have is the *apparatus criticus* – there for the qualified expert but not generally considered necessary for the non-expert or general reader – according to Wikipedia, the apparatus criticus is 'an organized system of notations to represent, in a single text, the complex history of that text in a concise form useful to diligent readers and scholars' – it contains, in other words, a great deal of information in highly condensed and coded form, the purpose of which is to show the author's grasp of the history of the transmission of often contested readings and to justify the same author's decision as to which reading he/she chooses to accept – there is then a decision made, with Roland Barthes' notion of the 'paradigm' in full operation – the 'paradigm' says Barthes, is 'the opposition of two virtual terms from which, in speaking, I actualize one to produce meaning…where there is paradigm (opposition), there is meaning → elliptically put: meaning rests on conflict (the choice of one term against another), & all conflict is generative of meaning: to choose *one* and refuse the *other* is always a sacrifice made to meaning, to produce meaning, to offer it to be consumed',[2] – it is, as Barthes puts it, a 'structural creation that would defeat, annul, or contradict the implacable binarism of the paradigm by means of a third term' –

the apparatus criticus is a detailed, forensic examination of the Greek terms, phrases, sentences, proposed to be those of Herakleitos – the intense letter-by-letter focus on the old Greek is visible almost everywhere in current pre-Socratic studies – but there's no corresponding laser-like interest in the word-by-word *English* into which these ancient Greek terms, which grew in a cultural soil very unlike our own, are regularly rendered – as if nothing is to be taken for granted in the Greek, but everything is to be taken for granted in the English of translation – as if the English into which we are translating simply 'comes naturally', needing no question, no examination, no caution as to where we are leading or misleading the ancient text – a perfect example is the almost universal translation of the Greek ψυχή as English 'soul' –

ψυχή appears ten times (including twice in one saying) in Herakleitos : D36, 45, 77, 85, 98, 107, 115, 117, 118 in Robinson's edition – the word has been rendered 'soul' in the translations of Robinson, Kahn, Patrick, Marcovich, Lebedev, Kirk and Raven, John Barnet, McKirahan, and Heidegger's German has come to 'soul' in English – the exception is Davenport who, no doubt fig-

uring there may be no exact English equivalent, plays it safe by using 'psy-
che' which nevertheless, thruout current English and American usage, has
its own problems –

Cunliffe on ψυχή : 'the animating principle, the vital spirit, the soul, the
life, life, one's life' > 'A disembodied spirit, a shade' – this latter meaning
is picked up in LSA as 'in Homer, only as *a departed soul, spirit, ghost*, which
still retained the shape of its living owner' – 'living owner'? – now, there's
a fine bit of loose-lipped language – if one's a shade, then one's no longer
living, and presumably the spirit cannot be said to own the departed body –
how one might own one's ψυχή is not explained, and I take it that if one is
dead, one does not own it any more – but Herakleitos is not much interested
(but for D98) in the fate or condition of ψυχή post-mortem – all his sayings
have to do with its condition in folk who are alive[3] –

only two sayings include the definite article with the noun : D98, with αἱ
ψυχαί (in the plural), & that they have the sense of smell in Hades (the one
time he refers to ψυχή after death – and D117, with τὴν ψυχήν, where the
article properly translates as 'his' rather than 'the' – so the notion of *the* soul,
as an entity distinct from other entities including the body, is not present in
Herakleitos – as thinking is common to all (D113) so is ψυχή, yet one's own
share is particular – Ezra Pound's almost shocking remark is more Herakleitean : 'That the body is inside the soul' – as we are both in and of ὁ λόγος –
recalling A N Whitehead's 'our individual share in the general implication' –

the word 'soul' derives, not from Greek or Latin, but from Old German,
Old Saxon, Old English – in SOED it means 'the principle of life in man or
animals; animate existence...the seat of the emotions, feelings, sentiments...
intellectual...power' – in Bailey's 1770 *Dictionary* it means : '...vital principle...Spirit, essence, principle...Interior power...qualities of the mind';
in Skeat's *Etymological Dictionary* it means : 'The seat of life and intellect in
man' – my quotes here are selective, yes, but serve to show how close many
historically verifiable meanings tally with the use Herakleitos makes of
ψυχή – other, later meanings that propose 'soul' as a kind of faculty, with a
distinct existence separate from the body, are mixed up with christian the-
ologies which have the soul as 'immortal' and what survives death to reside
in 'heaven' or in 'hell' post-mortem – recent Afro-American use of 'soul',

however, combining the notions of 'interior power' & 'seat of the emotions', continues the ancient sense of 'vital principle' or 'animating power' as something in which all can participate, something all living things can 'tap into' – when we take into account that ψυχή is said to be 'unlimited' (D45) and 'increases' (D115), it's a small step to understand that its scope of meaning extends beyond any individual instance – as Raymond Adolf Prier puts it : 'I prefer the translation of "life-force" for this phenomenon'[4]

♦   it is fairly easy to see, say, a religious attempt to replace an ancient secular text with one of its own – it is less easy to spot today's commentators attempting to replace an ancient text's concrete sayings and metaphors with abstract & interpretive concepts in the language-games of modern professional philosophy and of the history of philosophy –

how could one interpret the ancient texts without either replacing them with our own, or recontextualising them into a verbal ethos which was not that of the original – is it even possible to interpret ancient texts in such a way that we are returned to them, instead of reconfiguring them into the languages of contemporary scholarship – should it not be the role of current reading to re-enliven the original texts, to see and hear them in a renewed brightness and clarity, rather than smother them with our need to fill in the intervening centuries with the noise of our own, & entirely other, concerns – should not our efforts be to return our eyes and ears & thought 'To the words themselves!' – it's what we do with a Shakespeare sonnet, a poem by Emily Dickinson, an ode by Keats, a fragment by Sappho or Archilochos – why not with the sayings of Herakleitos – why can we not read these writings as if they are entirely contemporaneous with us, as if, even, we are able to utter them, simply, as our own in our own time – to digest the words, rather than to assess them as if the philosopher had handed them to us for marking – after all, they are here, in my case, in my own home, the books on my shelves, the words in my mouth, the sounds in my ears, at 8.08 am on a misty Melbourne Monday – there is no 'past' here other than unreliable memory, an old cracked recording, a fading photographic plate, a half-remembered old romantic thing – a heap of 'black bones' (Democritus) fought over by those who want to drape their own flesh over them – c'est moi!

♦   beginning with the wish for a decent chat with our elusive philosopher, I have instead run into a number of roadblocks to the meeting – two of these are my own : 1) I do not know enough of the corpus of ancient Greek writings to unthread (Barthes' term) the various references & allusions to them in Herakleitos's sayings, and 2) I simply don't have enough Greek, not even enough to read and understand everything I see in the full Liddell & Scott dictionary – if that was all that might prevent such a conversation, I'd be happy to offer 'guilty as charged' and leave it at that –

yet, for a non-specialist like me, the very body of published works by classicists and professional philosophers can present as formidable a resistance to understanding as my own expressed & acknowledged shortcomings – the way to understanding via modern works and editions, including the ways in which they are typically presented, is also the way of preventing understanding to occur outside of the academy – this is not to discredit anybody, or any edition, or any work, but I do think that the general reader (who is nevertheless invited to buy the books by their publishers – looking at any issue of *The New York Review of Books* or *London Review of Books* is all the research required) is rarely a factor in the preparation of modern classical texts by their modern authors – true, it is no longer usual for these books to present all their non-Greek text in Latin (tho I do have one edition, made in 1981, that does) – in a way, a prime excuse for what I'm writing here is that, in a more or less liberal more or less democracy like ours in the West, we general readers do have the right to speak of our actual experiences, including those we have when we read specialist texts that we are invited to buy –

in the editions I have, there are two main approaches to this matter – the first is, not to approach it at all, simply talk only to the specifically educated/qualified, provide a rigorous critical apparatus, & let a general reader cope as they may – the second, is to make an edition specifically for a general reader and simply leave the critical apparatus out – the problem with this second manoeuvre is that it has the same effect for the general reader as the edition that includes the critical apparatus – both editions exclude the general reader from the deep understandings gained by intensive scholarship by both the inclusion of the critical apparatus and by its exclusion –

a few works have tried to bridge the gap, Eva Brann's for example, and much commentary by Charles Kahn, T M Robinson, and the earlier John Burnet has made the attempt, but the gap they are looking at is not the one I'm looking at : their analyses have been primarily about the philosopher's *themes* rather than about the philosopher's *language* – it's not that I have a particular mechanism in mind, somewhere between the critical apparatus and the thematic essay, but rather a process whereby the depth provided by the critical apparatus might be expressed in plain language, the language of a general reader – else, the best I can do here is quote the physicist Werner Heisenberg : 'Even for the physicist the description in plain language will be a criterion of the degree of understanding that has been reached' –

roadblocks, however, are not all that I have met with along the trajectory of this undertaking – there has not only been a closer acquaintance with many volumes in my own library, but I have now come to a more open appreciation of the quantity & the quality of much commentary & analysis in their pages – even some of the most combative scholars have insights and an intellectual drive that I can admire but never emulate – the flow of books & articles does not stop, & the websites of The Center for Hellenic Studies & Academia.edu regularly attest to that – I have also found that information that may seem to have little direct relation to the field in question can usefully be brought to bear on any question, any question at all –

my amateur effort will continue, but I have no goal in mind, there's no sign ahead saying 'you have arrived', no point of destination – there is only the task, the journey, the attentiveness, the openness to whatever is absent so it may speak – New Zealand poet Allen Curnow once wrote of 'A small room with large windows'; Anne Carson wrote : 'writing involves some dashing back and forth between that darkening landscape where facticity is strewn and a windowless room cleared of everything I do not know'; artist Robert Motherwell, talking about his Open series of paintings, says they are windows, 'making the inside and the outside identical' – & yet I'd want, before all else, to resist any kind of rounding it off, there being no last knot to tie, as it were – no prospect, ever, of there being, for any of it, the last word

# notes

1 preamble

   1. Charles H Kahn, *The Art and Thought of Heraclitus*, Cambridge, UK, Cambridge University Press 1979, 116.

   2. Charles Olson, 'Maximus, to himself' in *The Maximus Poems*, London, Cape Goliard Press 1970, unpaginated.

   3. Charles H Kahn, ibid 29.

   4. Andrei Lebedev, *The Logos of Heraclitus: a Reconstruction of his Thought and Word (with a New Critical Edition of the Fragments)*, St. Petersburg, Nauka Publishers 2014. See https://www.academia.edu/8188629/Andrei_Lebedev_New_edition_of_Heraclitus_fragments_Greek_text_with_apparatus_criticus_and_English_translation

2 removing a distraction

   1. Juliet Fleming, *Cultural Graphology: Writing after Derrida*, Chicago, Chicago University Press 2016, 2.

   2. Random Cloud, 'from *Tranceformations in the Text of "Orlando Furioso"'* in *New Directions in Textual Studies*, eds Dave Oliphant and Robin Bradford, Austin, Harry Ransom Humanities Research Center, University of Texas at Austin 1990, 61–85.

   3. '...I call Neutral everything that baffles the paradigm...the opposition of two virtual terms from which, in speaking, I actualize one to produce meaning'. Roland Barthes, *The Neutral*, trans Rosalind E Krauss and Denis Hollier, New York, Columbia University Press 2005, 6–7.

   4. D F McKenzie, 'Speech-Manuscript-Print' in *New Directions in Textual Studies*, eds Dave Oliphant and Robin Bradford, Austin, Harry Ransom Humanities Research Center, The University of Texas at Austin 1990, 103.

3 kick-off

   1. Hans-Georg Gadamer, *Truth and Method*, trans rev Joel Weinsheimer and Donald G Marshall, London and New York, Bloomsbury 2013, xxviii.

   2. Ludwig Wittgenstein, *Zettel*, eds G E M Anscombe and G H von Wright, Berkeley and Los Angeles, University of California Press 1970, 2e.

   3. Martin Heidegger and Eugen Fink, *Heraclitus Seminar*, trans Charles H Seibert, Illinois, Northwestern University Press 1993, viii.

   4. Kathleen Freeman, *Ancilla to the Pre-Socratic Philosophers*, Oxford, Basil Blackwell 1956.

   5. Martin Heidegger, *Being and Time*, trans John Macquarrie and Edward Robinson, London, SCM Press 1962.

6. Martin Heidegger and Eugen Fink, ibid 26.

7. Ibid 3.

## 4 the unusual suspects

1. T M Robinson, *Heraclitus: Fragments*, Toronto, University of Toronto Press 1987, 195–200.

## 5 the reliably unreliable text

1. M Marcovich, *Heraclitus, Greek Text with a Short Commentary*, Venezuela, Los Andes University Press 1967, 163–164.

2. Ingram Bywater, *Heraclitus of Ephesus*, trans G T W Patrick, Chicago, Argonaut Inc 1969, 123.

3. Guy Davenport, *Herakleitos*, Berkeley, Peter Koch 1990.

4. Robert Bringhurst, *The Fragments of Parmenides* (wood engravings by Richard Wagener), Berkeley, Editions Koch 2003.

5. Parmenides, *ΠΕΡΙ ΦΥΣΕΩΣ*, French translation by Jean Zafiropulo, Paris, Alberto Tallone 1953.

## 8 in other words / in no other words

1. Rochelle Altman, *Absent Voices : The Story of Writing Systems in the West*, Delaware, Oak Knoll Press 2004, 5–8.

2. Eva Brann, *The Logos of Heraclitus*, Philadelphia, Paul Dry Books 2011, 4.

3. Ludwig Wittgenstein, *Zettel*, eds G E M Anscombe and G H von Wright, Berkeley and Los Angeles, University of California Press 1970, 28e.

4. Lewis Hyde with painter Max Gimblett, *The Ten Oxherding Pictures*, an exhibition first held at The Japan Society, New York, 2010. See http://www.lewishyde.com/archives/oxherding

5. Robert Bringhurst in *Carving the Elements: a Companion to The Fragments of Parmenides*, Berkeley, Editions Koch 2004, 12.

6. Martin Heidegger, *Heraclitus*, trans Julia Goesser Assaiante and S Montgomery Ewegen, London and New York, Bloomsbury 2019, xviii.

## 9 stray notes & reminders

1. Lewis Carroll, 'Through the Looking Glass and what Alice found there', in *The Illustrated Lewis Carroll*, ed Roy Gasson, Dorset, New Orchard Editions Ltd, undated, 168.

2. Walter Burkert, *The Derveni Papyrus on Heraclitus (Col. IV)*, Zurich Open Repository and Archive, University of Zurich 2011, https://www.zora.uzh.ch

3. Andrei Lebedev, 'The Derveni Papyrus and Prodicus of Ceos (dedicated "To the memory of Martin West") 2019' in *Indo-European Linguistics and Classical Philology – XXII*, Proceedings of the 22nd Conference in Memory of Professor Joseph M. Tronsky – June 18–20, 2018.

10 is it a poem / is it a prose | is it a writing / is it a saying

1. Wilfred E Major and Michael Laughy, *Ancient Greek for Everyone*, online course at https://ancientgreek.pressbooks.com/front-matter/introduction/
2. Eva Brann, *The Logos of Heraclitus*, Philadelphia, Paul Dry Books 2011, 4.
3. Ron Silliman, *The New Sentence*, New York, Roof Books 4th printing 2003.
4. Robert Bringhurst, 'Raven's Wine Cup' in *Carving the Elements*, Berkeley, Editions Koch 2004, 136.
5. James Reeves, *A Short History of English Poetry*, London, Heinemann 1961, xv.
6. Eric A Havelock, 'Pre-literacy and the Pre-Socratics' in *Bulletin of the Institute of Classical Studies*, Vol 13, Issue 1, December 1966, 55.
7. R P Austin, *The Stoichedon Style in Greek Inscriptions*, New York, Arno Press 1973, 18.
8. Eric A Havelock, ibid, 57.
9. Ibid, 66.
10. Charles Olson, 'The Advantage of Literacy is That Words Can Be on the Page' in *Collected Prose*, eds Donald Allen and Benjamin Friedlander, California, University of California Press 1997, 353.
11. Paul Saenger, *Space Between Words: The Origins of Silent Reading*, California, Stanford University Press 1997, 244–245.
12. *Letters of John Keats*, A new selection by Robert Gittings, Oxford, Oxford University Press 1970, 43.

11 to the alphabet & back

1. Stephen V Tracy, *Attic Letter-Cutters of 229 to 86 B.C.*, California, University of California Press 1990.
2. Stephen V Tracy, email conversation with the author 20 May 2021.
3. R P Austin, *The Stoichedon Style in Greek Inscriptions*, New York, Arno Press 1973.
4. L H Jeffrey, *The Local Scripts of Archaic Greece*, rev ed by A W Johnston, Oxford, Oxford University Press 2003.

12 the λόγος files

1. Walter Burkert, *The Derveni Papyrus on Heraclitus (Col. IV)*, Zurich Open Repository and Archive, University of Zurich 2011, https://www.zora.uzh.ch
2. Aulus Gellius, *Attic Nights*, trans John C Rolfe, Loeb Classical Library, 3 Vols, Cambridge, Harvard University Press reprinted 1946.
3. Leonardo da Vinci, *The Notebooks of Leonardo da Vinci*, trans Edward McCurdy, 3 Vols, London, The Reprint Society 1954.
4. Joseph Joubert, 'The Notebooks of Joseph Joubert', in *Translations*, trans Paul Auster, New York, Marsilio Publishers : EW Books 1997, 21–166.
5. Louis Zukofsky, 'A-12' in *A*, California, University of California Press 1978.

6. Alan Loney, *Crankhandle : Notebooks November 2010 – June 2012,* Melbourne, Cordite Books 2015.
7. As it happens, there is a digital typeface that can be employed, Diogenes, designed by Christopher Stinehour, and done for an edition of Parmenides, translated by Robert Bringhurst, printed and published by Editions Koch in Berkeley California in 2004 – Stinehour's Diogenes has not yet been made commercially available, but as a digital type it strikes me as perfect for being able to give any reader, expert or general, a taste of what the Derveni Papyrus lettering might look like – the type is all in capital letters, based on 6th century BCE inscriptional lettering.
8. Gregory Nagy, 'Traces of an Ancient System of Reading Homeric Verse in the Venetus A' in *Recapturing a Homeric Legacy : Images and Insights From the Venetus A Manuscript of the Iliad,* Hellenic Studies 35, see https://chs.harvard.edu/wp-content/uploads/2020/11/Recapturing_Homeric_Legacy.pdf

## 13 refusal of the last word

1. Andrei Lebedev, *The Logos of Heraclitus: a Reconstruction of his Thought and Word (with a New Critical Edition of the Fragments),* St. Petersburg, Nauka Publishers 2014. See https://www.academia.edu/8188629/Andrei_Lebedev_New_edition_of_Heraclitus_fragments_Greek_text_with_apparatus_criticus_and_English_translation
2. Roland Barthes, *The Neutral,* trans. Rosalind E Krauss and Denis Hollier, New York, Columbia University Press 2005, 7.
3. Raymond Adolf Prier, *Archaic Logic: Symbol and Structure in Heraclitus, Parmenides, and Empedocles,* The Hague, Paris, Mouton 1976, 72.
4. It is, however, important to note that 'Followers of Orphic and Pythagorean sects believed their souls had a distinct existence in Elysium' (Edward Jenner, personal communication with the author). In Jenner's account of 'pre-Christian ideas about the soul' (*The Gold Leaves,* Atuanui Press, Auckland 2014) he writes: 'These minute shreds of gold leaf, placed in the mouths or the hands of dead initiates, bear messages that gave their owners a supreme confidence in the survival of the soul and its progression towards an afterlife in a realm of sunlit groves and meadows' (p9). The question whether Herakleitos accepts the survival of ψυχή as the gold leaves propose may depend on whether one accepts the Herakleitos references as being positively literal or negatively ironic.

www.ingramcontent.com/pod-product-compliance
Lightning Source LLC
Chambersburg PA
CBHW051845040426
42447CB00006B/713